THE INSECURE CITY

THE INSECURE CITY

Space, Power, and Mobility in Beirut

KRISTIN V. MONROE

RUTGERS UNIVERSITY PRESS
New Brunswick, New Jersey, and London

Library of Congress Cataloging-in-Publication Data

Monroe, Kristin V., 1974– author.
 The insecure city : space, power, and mobility in Beirut / Kristin V. Monroe.
    pages cm
 Includes bibliographical references and index.
 ISBN 978–0–8135–7463–9 (hardcover : alk. paper) — ISBN 978–0–8135–7462–2 (pbk. : alk. paper) — ISBN 978–0–8135–7464–6 (e-book (epub)) — ISBN 978–0–8135–7465–3 (e-book (web pdf))
  1. Sociology, Urban—Lebanon—Beirut. 2. Public spaces—Lebanon—Beirut. 3. City traffic—Lebanon—Beirut. 4. Violence—Lebanon—Beirut. 5. Urban anthropology—Lebanon—Beirut. 6. Beirut (Lebanon)—Social conditions. I. Title.
 HT147.L4M66 2016
 307.76095692′5—dc23

                                                                                            2015021869

A British Cataloging-in-Publication record for this book is available from the British Library.

Copyright © 2016 by Kristin V. Monroe
All rights reserved
No part of this book may be reproduced or utilized in any form or by any means, electronic or mechanical, or by any information storage and retrieval system, without written permission from the publisher. Please contact Rutgers University Press, 106 Somerset Street, New Brunswick, NJ 08901. The only exception to this prohibition is "fair use" as defined by U.S. copyright law.

Visit our website: http://rutgerspress.rutgers.edu

Manufactured in the United States of America

For my mother, Ann

CONTENTS

	List of Figures	ix
	Acknowledgments	xi
	Note on Language	xv
	Introduction	1
1	The Privatized City	18
2	The Space of War	35
3	Politics and Public Space	56
4	Securing Beirut	79
5	The Chaos of Driving	101
6	"There Is No State"	121
	Conclusion	139
	Notes	145
	References	165
	Index	177

FIGURES

Figure I.1	Internal Security Forces billboard	15
Figure 1.1	Map of Beirut in Lebanon and the region	19
Figure 1.2	Neighborhood map of Beirut	20
Figure 1.3	Cafe at a Beirut public garden during the late Ottoman period, ca. 1900–1920	23
Figure 1.4	The Corniche	32
Figure 2.1	Tent City	52
Figure 3.1	Lebanese "Parties and Colors" albums and stickers	61
Figure 3.2	Traffic sign: "No foreign intervention! 1559 prohibited from passing"	77
Figure 4.1	Shark-fin barriers	84

ACKNOWLEDGMENTS

Before reaching this point, I had heard it said countless times that writing a book is a journey, and indeed it has been. I could not have done it alone. I am deeply indebted to the residents of Beirut who shared their time, energy, and stories with me. While some of those I knew and spoke with are mentioned by name in this book, most are not. To all of you, I express my gratitude for your friendship, assistance, and insights.

Throughout the process of developing this book, numerous colleagues at the University of Kentucky provided feedback and support. These include: Karen Rignall, Srimati Basu, Patricia Ehrkamp, Janice Fernheimer, and Carmen Martinez Novo. Sarah Lyon and Mark Whitaker have offered valuable guidance throughout the various stages of developing the book, and I am very grateful for their time, energy, and interest in my work. Many others, including Shannon Bell, Jacqueline Couti, Nazera Wright, Cristina Alcalde, Scott Hutson, Shannon Plank, Hang Nguyen, Paul Chamberlin, Jim Ridolfo, Erin Koch, Lisa Cliggett, and Diane King, offered their support in other ways, most especially by providing me with a community of scholars and friends with whom I could exchange ideas and, importantly, laugh. I thank all my colleagues in the Department of Anthropology at the University of Kentucky for their unwavering collegiality and good cheer.

I am grateful to the American University of Beirut's Center for Arab and Middle Eastern Studies for providing me with an institutional home during my first research period. Kirsten Scheid and Joy Farmer were exceedingly helpful and wise colleagues and friends during my fieldwork. Mona Harb and Mona Fawaz were always encouraging. I thank Jehan Mullin and Mariko Shimomura for their friendship during those tumultuous years in Beirut. At Northwestern University, I benefitted greatly from the mentorship of Micaela di Leonardo and Jessica Winegar and the support of the anthropology department as a whole. Nicole Fabricant has been a close and enthusiastic reader of drafts and has helped me develop my thinking. In different ways and at various points, Farha Ghannam, Setha Low, Aseel Sawalha, Beth Notar, Julie Peteet, and Lara Deeb have offered encouragement that has sustained me through this project.

Several individuals were instrumental in forging my path as a scholar including Bill Hoynes at Vassar College, and, later, Nina Berman and Denise Spellberg at the University of Texas at Austin helped to shape my theoretical outlook and historical perspective. At Stanford, I benefitted from the support and guidance of Sylvia Yanagisako and James Ferguson. Their inquiry about our social world, theoretical interrogation, and analytical rigor are qualities I attempt to model in my own pursuits. I learned invaluable lessons in ethnographic writing and analysis from Renato Rosaldo. I thank Shelly Coughlan and Ellen Christensen for their assistance, and I am also grateful for the colleagueship of Tania Ahmad, Oded Korczyn, Zhanara Nauruzbayeva, Yoon-Jung Lee, Mun Young Cho, Jocelyn Chua, Sima Shakhsari, Aisha Beliso-De Jesus, and Kutraluk Bolton. Tiffany Romain's incisive comments and suggestions made me a better writer and her calm spirit gave me ballast. Mukta Sharangpani was always a beacon and gave me a sense of both home and sisterhood during those years.

The research from which this book emerges would not have been possible without the generous financial support of several institutions and agencies. These include Stanford University's Department of Anthropology and the Freeman Spogli Institute for International Studies, a Fulbright-Hays Doctoral Dissertation Research Abroad grant (2004–2005), a Geballe Dissertation Fellowship from the Stanford Humanities Center, and a summer research grant (2010) from Northwestern University's Dispute Resolution Research Center at the Kellogg School of Management. The University of Kentucky College of Arts and Science's start-up funding for new faculty enabled me to conduct follow-up research in summer 2013; the Woodrow Wilson National Fellowship Foundation's Career Enhancement Fellowship allowed me to focus full-time on writing during the 2013–2014 academic year; and a College Research Activity Award from the University of Kentucky's College of Arts and Science provided support for the book's production.

I would also like to thank my editor, Marlie Wasserman, for her support of this book project, and the staff at Rutgers University Press for their assistance throughout the publishing process.

Finally, I wish to thank all my friends and family, without whom I could not have accomplished this goal. Annette Muller's enthusiasm for and faith in my work and thinking have been a source of inspiration over the last decade. During a late night conversation outside her San Francisco apartment long ago, Robin Li set me on this path, and it is one that I could not have imagined for myself. I am always grateful for her wisdom, warmth, and, imagination. This book is dedicated to my mother, Ann Monroe. In this venture, as in all others I have undertaken in life, she has offered boundless support and love. Her belief in me has motivated my work and given me the confidence to travel to new places and take on new challenges. I thank Louay Faissal for his patience, endurance, and assistance in making this idea of writing and finishing a book a reality. This book bears witness to a journey that runs parallel to this project, the one—from Beirut to Lexington, Kentucky, and all the stops along the way—that brought us together. The light and laughter of our daughter Maysan and the arrival of Noor have made the final stages of this journey more beautiful, engaging, and fun.

NOTE ON LANGUAGE

People I met in Beirut spoke mainly in Lebanese dialect, English, and French and very often a mix of languages. I have used an extremely simplified system for transliterating the Lebanese Arabic dialect, omitting all indication of long versus short vowels as well as distinctions between hard and soft letters. I trust that specialist readers will be able to use the context to follow my transliteration. Names of places and people adhere to their official or common spellings. All translations of Arabic- and French-language textual materials are my own unless otherwise noted.

THE INSECURE CITY

INTRODUCTION

AN AMERICAN IN BEIRUT

My first knowledge of Beirut came from television news about Terry Anderson, a U.S. hostage captured during the Lebanese civil and regional war by Hizbullah militants in 1985 and held in captivity for six years.[1] Anderson grew up in a town near where I spent my childhood and local news coverage during the years of his captivity regularly featured members of his family, most especially his sister and her efforts to gain his release. Thus, my first image of Beirut was one of war. Years later, when I was a graduate student, a professor, knowing of my interest in issues of class and urban space in the Middle East, suggested I visit Beirut. After a preliminary visit in 2003, I was struck, as any visitor is, by Beirut's vibrancy and diversity, the coexistence of so many different ways of living: peddlers with carts overloaded with seasonal produce sharing the street with global corporate retail outlets; a woman in conservative Islamic dress having coffee with a friend wearing a revealing outfit; sleek high-rise residential buildings being constructed alongside timeworn two-story houses. But I was also struck by the class and status aspects of urban public life and culture and was surprised to find that studies of class in Lebanon were relatively few as issues related to political sectarianism have long been the primary subject of scholarly inquiry. From the start, then, even as Lebanese told me "you'll never figure out how social class works here" or "we don't really have social class here," I set out to explore how class and status mattered in the space of Beirut. As I describe later, my further focus on mobility, as a particular way of using space, came from living in the city and the concerns of people I met.

A host of sad and disruptive events, which I explore more fully throughout the book, unfolded during my research, which was undertaken in three phases: October 2004 to June 2006, summer 2010, and June 2013. My first and most extended period of research coincided with the initial stages of violent political unrest that began just prior to and in the wake of the assassination of former prime minister Rafiq Hariri in early 2005. These were difficult times. Nearly every month, a political figure or journalist was assassinated by a car bomb. Other bombs exploded in commercial or industrial locations of the predominantly Christian parts of Beirut and its suburbs. Security measures and blockades sprung up near anticipated targets. Schools were closed an inordinate number of days in 2005. Apart from the optimism that existed for those allied with the anti-Syrian March 14th political coalition upon the full withdrawal of the Syrian army from Lebanese land in April 2005, the overall mood in the city that year was both depressed and anxious. In early 2006, intersectarian tension was on the rise, and it flared up in Beirut during the controversy over the publication of cartoons satirizing the Prophet Mohammad by a Danish newspaper. And, just a few weeks after my return to the United States in mid-June 2006, times grew far worse. In July 2006, in retaliation for Hizbullah's capture of two Israeli soldiers, Israeli bombs began to land first in southern Lebanon and then in South Beirut and elsewhere around the country. I watched the images from the United States, from safety, and read the news and e-mail missives with apprehension. I learned that the apartment where I had lived was now a temporary refuge—opened by a friend and the new tenant— for a Palestinian family fleeing their home, which was located near Dahiya (literally "the suburb" in Arabic but used to refer to the southern suburbs of Beirut), the area under heaviest assault. By the war's end in August 2006, immense infrastructural damage blighted the entire country, but most extensively Dahiya and southern Lebanon. Thousands of people were displaced. The Israeli Defense Force's indiscriminate ground and air strikes, according to Human Rights Watch, resulted in the deaths of approximately 900 civilians.[2]

It would be difficult for me to measure, or to isolate, the ways in which these events shaped my research. They shaped the project completely. Closures in and of the city, which occurred following an explosion or had to do with security measures, were the cause for the delay or cancellation of interview appointments. In the weeks following Hariri's assassination, I was cut off from any formal or informal research activities. Lebanese were fearful about what violence would occur next, much of the nation was in mourning, and the practices of everyday life came to a halt. And then regular life would begin again, but Beirutis, I quickly learned, resumed their activities with an awareness that beneath the surface of normality lay the possibility for everything to come apart again. This kind of cycle, of violence-stop-pause-resumption, punctuated the months of 2005 and 2006 after an assassination or bombing attack occurred. Between the bombs, there existed what anthropologists Roma Chatterji and Deepak Mehta (2007) call "the recovery of the everyday." Anxiety, born from anticipation about what might happen next, resided in this everyday. During this period in Beirut, there were moments when the agenda of research was neither a practical choice nor a compassionate one.

The bombs, and the anxiety, did not stop people from living, of course. And thus, while the first period of my research showed me a Beirut in distress, it also cheered me with its warmth, its humor, and its energy. In juxtaposition to the threat posed to the public and the dejection felt by Beirutis at the return of violence and political sectarian strife to their city's streets, scenes of everyday public sociality showed another side. This is the side of Beirut that sociologist Anthony Giddens (1984) would describe using the term *co-presence*, or face-to-face interaction. It is an aspect of Beirut's public life that goes beyond the trope of Lebanese as resilient in the face of challenges and that counters Richard Sennett's (1974) description of the isolating, "stale and empty" public that has come to characterize the modern, Western city. More specifically, it is the side of Beirut that includes a descending basket on its way from an upper floor of an apartment building to the market on the ground floor. An apartment dweller leans out over the veranda and calls down to the store worker a list of things needed for a recipe or meal already in progress. The items are placed in the basket, and it goes up to the veranda. Money is then put in the basket, and it is sent back down to the store. Everyone is satisfied. Moments like these, those I only observed and those that I also took part in—for instance, the paying of all my bills in person to employees from the electricity, internet, and water companies who came to the door of my apartment—are not only charming to the outsider but seem to serve as a kind of salve for Beirutis living with the history, presence, and anticipation of conflict and divides among its people.

The stops and starts, the horrific spectacle and aftermath of bomb explosions, and the return of political sectarian strife to the public realm could be described as challenges to the conduct of research. At the same time, however, such a description would be ill-fitting in the sense that these events were experiences that came to constitute the research itself. And, often, challenges having little to do with the political crises appeared more formidable. For example, the cultural capital I possessed, initially as "a researcher from Stanford," held sway only to a limited degree in many social and professional settings in Beirut. Another kind of capital, which is issued in the form of connections or relations to a particular person, was usually valued much more by people with whom I sought to set up meetings and interviews. Getting my foot in the door usually required being able to mention that I was in some way connected to a person whom a potential interviewee knew and trusted. Some interviews, like the one I conducted with the head of the traffic-police division, required weeks of advance effort, during which I met with people sequentially. One person would bestow on me access to the next and so forth until I had worked my way up the chain of command. This is also the process by which I secured interviews at Solidere, the corporation responsible for rebuilding the downtown area. This practice of establishing and finessing connections or favors through face-to-face interaction is also a topic I take up in the book with regard to people's sentiments about everyday forms of corruption. "Knowing someone who knows someone" was often my only means of gaining access to professionals such as architects, engineers, academics, representatives of nongovernmental organizations (NGOs), and government officials. While this description may fit any number of research locales, Lebanon's relations of patronage, which are often mobilized to circumvent state authority, heightened the necessity of being tied to and circulating through stratified networks of influence.

Outside the professional arena, with respect to my entrée into Lebanese society more generally, I encountered similar, though less tangible kinds of limitations that too shaped my research. First, through formal tutoring arrangements and informal viewing of Lebanese television programs and everyday conversations, I had to "Lebanize" the formal Arabic I had studied and spoken for years in classrooms. Second, the fact that I chose to live alone and had neither real nor fictive ties to a Lebanese family meant that I was fairly unmoored in a society where family life is the basis of the social fabric. On Sundays, when Beirutis spend the day with their extended families and most commercial life shuts down, being out in the city was a lonely, traffic-free excursion. All of us from outside Lebanon—the foreign migrant workers, ex-pats, students, and researchers—we the familyless would encounter one another on the near-empty streets. I also found developing relationships with Beirutis from outside the middle class to be a challenge. While I have a working-class background, my level of education, the fact that I was a foreign student/researcher living outside my home country, and the way that I could afford to live as middle class in Beirut situated me squarely in a middle-class world. I took steps to try to expand this world by, for example, doing volunteer work with two different organizations working with underprivileged Lebanese youth, volunteering with an agency helping process paperwork for Sri Lankan workers trying to get home after the Asian tsunami in January 2005, and trying actively to forge research relationships with working-class residents of the city through acquaintances and contacts.

Alongside these limitations, however, I also experienced a certain kind of public access. As a mixed-race African American woman, I did not stand out when walking down the street in Beirut. Before I spoke, I was assumed by Lebanese to be of North African heritage but possibly part Lebanese, as I would come to learn through countless conversations that ensued after I began speaking my non-native Arabic in taxis, stores, restaurants, offices, and the like. Blending into the landscape of the city enabled me to move around without being outwardly perceived as being from outside the Arab world. I could therefore travel without being immediately marked as a Westerner, as many other researchers and visitors from the West typically are. In parts of Dahiya, Beirut's southern suburbs largely secured by Hizbullah, the way I look afforded me with a kind of right of entry that was often denied to other Westerners identified as such by Hizbullah security members positioned on the streets.

Another kind of access came in a different form. While I was foreign and Western, a perceived ally perhaps of the policies pursued in the region and Lebanon by the U.S. government, I did not possess a political or sectarian affiliation and my history came from elsewhere. In this sense, not being Lebanese endowed me with a certain kind of perceptual and geographic privilege to draw my own cartographies of the city and nation. As they were not historically constituted through experiences of violence and fear, my mappings were necessarily distinct from those that might be drawn by a native of Lebanon, a survivor of the long civil and regional war, or a person whose home had been destroyed by Israeli bombs. For me, the terrain was more open. As a foreigner and outsider, I was also furnished with a look in. I was, time and again, told by Lebanese how and who the Lebanese are: that is, "all Lebanese are like X" or "the Lebanese don't care about X," or "X behavior is typical Lebanese."

In addition to observing and participating in public space and in informal conversations in taxis, buses, and on the streets, I conducted interviews with a diverse group of residents of the city, including foreign workers, as well as representatives from both state and nongovernmental agencies and institutions involved with urban planning, traffic safety and enforcement, and civil society. I also gathered and analyzed historical materials from newspapers housed in the libraries of the American University of Beirut and Saint Joseph's University, the Lebanese National Archives, the Centre d'Etudes et le Recherche sur le Moyen-Orient, and at the archives of the newspapers *An-Nahar* and *L'Orient le Jour*.

"YOU HAVE TO LOOK MORE BROADLY AT THE ISSUE OF TRAFFIC"

During a conversation I was having in 2005 with Reem, a woman in her early forties who worked as an administrator at a university in the northern Beirut suburb of Louaize, we were talking about the hassles of living in Beirut. It was the topic of traffic that set her off. "Look," she interrupted when I began to ask her about the new traffic lights being put up in the city, "you have to look more broadly at the issue of traffic; it tells you a lot about what is happening in our society. People are stressed—this is what traffic is about; people are oppressed and we have been living in a police state since the end of the [civil] war. It's a bad economic situation, taxes, education is expensive, food is expensive. . . . What you see on the roads? You see that people are fighting for space. They don't have space at home, the economy is bad, and there are no forests, no parks, no places to breathe. There is only concrete. . . . Four times this week I almost got killed driving. Finally, I had a meeting in Broumanna and I took a taxi there. I was not going to drive myself."

When Reem explained in frustration how traffic problems were linked with, as she put it, "what was happening in society," it was one of many similar moments during my research: when people talked to each other and me about getting around Beirut, they were also expressing larger concerns about social, political, and economic life. Talk about mobility experiences, in fact, exposed some of the inequalities of the city itself.[3] Stories told among passengers in shared *service* (French pronunciation) taxis[4]—the city's most widely used form of public transportation—on television programs, on stage, and through popular culture told of how getting around Beirut was about more than merely getting somewhere;[5] it was about how people encountered the very formation of their civic culture in a city wounded by war and, once again, on the razor's edge.

Through stories and practices of daily life, this book examines how people's movements through a city are profoundly shaped by the insecurity of their lives. In 2004, fifteen years after the end of a protracted war in Lebanon (1975–1990), bombs reappeared in Beirut.[6] In September 2004, the Lebanese parliament voted to amend the constitution to allow for an extension of the presidential term. This amendment kept the pro-Syrian President Emile Lahoud in power for several more years and was a move thought to be orchestrated by the Syrian regime. To protest Lahoud's term extension, government ministers, including the prime minister, Hariri, resigned. One of these ministers survived an assassination attempt by car bomb in October 2004, and a few months later, on February 14, 2005, Hariri and twenty-one others in his motorcade were killed in a massive car-bomb explosion along a seaside road. A climate of assumed Syrian culpability for Hariri's death ensued, and tensions boiled over as Lebanese took to the streets demanding the withdrawal of the Syrian army from Lebanese territory and an end to Syrian involvement in Lebanese affairs. In the months after Hariri's assassination, two broad political coalitions, named after the dates of their mass protests in downtown Beirut, emerged: the March 8th group (a coalition of parties with both Shi'i Muslim and Christian affiliation) was aligned with the Syrian regime; the March 14th alliance (a group of parties with mainly Sunni Muslim and Christian affiliation backed by the United States and its regional allies) took an anti-Syrian-government stance. In the months to come, these political camps would become the two central political actors in what is referred to as the post-Hariri era in Lebanon.

Since these events, amid the changing course of Lebanese politics, residents have once more had to map out lives amid a contentious, often violent, political-economic landscape. In this book, I explore how experiences of moving through Beirut are characterized by a precariousness wrought not only by the anticipation of violence but also by the workings of class, political, and state power. In keeping with urban anthropology's longstanding attention to the relationship between urban space and social inequality, I examine how understandings and practices of spatial mobility in the city do more than simply reflect social differences;[7] they are also a means through which an uneven and insecure urban citizenship is produced as dimensions of hierarchy and power shape people's access to and experiences of urban space.

I arrived in Beirut in fall 2004. I expected to learn about the class dimensions of postwar reconstruction processes. But soon, after the first bomb targeting a political figure exploded, I realized that I was witnessing a renewal of violence that would challenge the notion of Lebanon as having moved past war altogether. From late 2004 to early 2006, what I found was not a long-term sustained war. But it was not peace either. It was something else, a place reorganized into divided parts by the resurgence of political sectarianism and the threat of bombs, a time of frustration and disaffection with the state, the economy, and the political order.[8]

In July 2006, a month-long full-scale war between Hizbullah and Israel broke out following Hizbullah's capture of two Israeli soldiers near the border.[9] In 2010, and again in 2013, I returned to Beirut to examine how stories about and practices of urban mobility captured the meanings of everyday life in this insecure city. I watched Beirutis of diverse backgrounds and perspectives move through the city and listened to their talk about what their journeys revealed to them and what these journeys revealed about Lebanon. These observations and stories form the book's ethnographic core.

The Insecure City

Many studies have pointed out how the organization of urban space in Beirut is part and parcel of the exercise of power and hierarchy. In her ethnography *Reconstructing Beirut: Memory and Space in a Postwar City* (2010), Aseel Sawalha looks at how ordinary residents there have been displaced by the city's post–civil war reconstruction. She highlights how the city has been rebuilt in ways that exclude poor and working people and the increasing takeover of urban space by political and economic elites. Like Sawalha in her concern with the spatial relations of power, Lucia Volk (2010) investigates Beirut's and Lebanon's many public memorials and shows how they express elite political authority in public spaces and the long-standing politics of cross-sectarian community solidarity. Sune Haugbolle (2010) also considers space and memory as he examines debates around the cultural production of memories about the civil and regional war in the years between 1990 and 2005 and shows how the war was at one and the same time represented and rendered invisible in Beirut's public spaces. Lara Deeb and Mona Harb (2013), in their study of how pious young Shi'i Muslims navigate leisure opportunities amid Beirut's sectarian and class hierarchies, look closely at the relationship between space and morality by examining how leisure sites such as cafes, gyms, and weddings shape young people's social behavior and practices. Scholarship about Beirut's geography has also been concerned with the role urban infrastructure and planning policies have played in producing spatial and social inequalities (Fawaz 2009b, 2009c) and the impact of conflict on urban life and politics (Fregonese 2009, 2012). All this impressive scholarly work offers incisive and multifaceted analyses of the spatial and social relations of power in Beirut, but it does not address the experiences of insecurity and security that were crucial aspects of everyday life during the time of my ethnographic research, when bombs were going off and the threat of political violence intersected with broader anxieties about living in uncertain and unprotected times.

Hence, this book aims to contribute to these studies of Beirut's landscape and those focused on other cities in the Arab world by addressing the relationship between urban space and the meanings and experiences of security and insecurity that emerge from people's movements through the streets.[10] By focusing on a wounded city shaped by its history and ongoing experience of conflict,[11] this book is part of a growing body of anthropological work exploring "how ethnographic subjects contend with matters of security and insecurity as they attempt to forge a life in a complex, conflictive, and often violent and dangerous social and political-economic milieu" (Goldstein 2010, 489). Like this scholarship, my book moves beyond traditionalist approaches to security that focus primarily on how it is fashioned at the level of policy and strategy making in government and military domains and toward a consideration of the lived encounter with security states and insecurity and, in this case, how these are reflected in movement in and around Beirut.

To analyze the situation at this level, I focus on three dimensions of insecurity. First, I examine the kinds of insecurities that surround everyday life in zones of conflict. In Beirut, as elsewhere in the Lebanon-Syria-Israel-Palestine region, this context of conflict is the outcome of a fractious political climate whose tenor is shaped by the broader field of regional geopolitics and its array of transnational actors. The periodic ebbs and flows of tension that characterize these zones of conflict require residents to stay on guard, always ready to adjust their present lives and daily course of activity, as well as to manage their fears about the future possibility of a protracted and full-scale war, in response to heightened tensions or signs of an oncoming crisis.[12] In getting around the city, Beirut's residents, in other words, are always aware that just beneath the surface of a normal day is the possibility for a violent disruption of everyday life that could last hours, a week, months, or more.

Second, I provide a picture of how the project of security can create insecurity in the pursuit of various goals, from the protection of certain people and the enactment of justice to the enforcement of boundaries that keep out "unwanted" or "undesirable" populations.[13] In the case of Beirut, I explore this problem by looking at the following paradox: In Beirut, the intensification of security by the state for "the few"—an overlapping group of class and political elites—through the installation of checkpoints and barriers and the rerouting of traffic in ways that close off not just street blocks but whole neighborhoods, requires the deployment of certain kinds of spatial and social practices on the part of "the many," whose lives, as a result, become disorderly, and, in this way, less secure. For instance, following former prime minister Hariri's assassination in February 2005, traffic on a key artery was changed from two-way to one-way as part of the fortification of security around the Hariri palace in Koreitem, my neighborhood in West Beirut.

Third, a sense of vulnerability that is both physical and existential is contained in my use of the terms *insecurity* and *unsafety*. Feelings of social and psychic unsafety that emerge in people's talk about being mobile in Beirut lead to blaming the state for being ineffective, corrupt, and forcing citizens to fend for themselves. These sentiments about not being taken care of by the state are of a piece with the kinds of insecurities anthropologists have described as a feature of the global human condition in the contemporary era, which is marked by neoliberal polices and governance involving the reduction of social safety nets and increased privatization, open markets, and deregulation. This is an era of downward socioeconomic mobility and anxious citizens who, feeling under threat and unprotected by the state, often move to take responsibility for their own security.[14] In using the term *unsafety*, I mean to convey a notion of insecurity or threat that includes but does not hinge on the possibility of bodily harm or injury, as the term *danger* often does.

Theories of security and insecurity have helped us understand the ways in which contemporary human life is besieged by a whole host of challenges and fears, from imminent ecological or financial disaster to the avoidance of toxins and crime, that make the management of risk and insecurity a central feature of our lives. While the generalizability of the concept of insecurity is useful insofar as it directs our attention to the significance of various social, environmental, and political phenomena, my ethnography shows how different kinds of insecurity converge in everyday experience. By considering how unsafety for Beirut's residents is engendered by various kinds of threats, from the dangers of political instability to the risks posed by reckless drivers and the moral and civic injuries of elite corruption, I highlight the tensions created by and the links among various forms of insecurity.

By engaging a notion of insecurity that attends to the ways in which people are rendered vulnerable, not only physically but also in a political, economic, and ontological sense, I am providing a multifaceted analysis of what insecurity means for residents of Beirut. Not only the possibility of political violence threatens Beirut's ordinary residents but also economic and class dynamics that make life precarious for ordinary folk amid weakened structures of social and economic support and rising inequality.[15] In this way, my account shows how the insecurities of urban and public life in Beirut are both an outcome of Lebanon's contentious geopolitical milieu and part of a broader global experience of downward socioeconomic mobility and common anxieties that characterizes the human condition in the early twenty-first century. Thus, although many of the experiences I describe in this book that relate to the militarized and precarious urban setting in which Beirutis live may seem quite distant from life elsewhere, in places in the rest of the world where there is peace and wars are not fought, for example, many other issues that people described to me are increasingly relevant: living with an intensified police and security presence, finding it ever more difficult to make ends meet, and coping with stress about the lack of a social safety net.

Mobility and Urban Space

My investigation of the meanings of security and insecurity in Beirut took place in the streets of the city. Being mobile in an urban environment, I propose, is a window into the everyday experience of civic life and citizenship and sheds light on how class, politics, and state power are spatialized in a place and time fraught with conflict and uncertainty. In many parts of the Middle East, conditions of insecurity that are an outcome of political violence, the practices of the security state, and the militarization of everyday life shape people's everyday mobility experiences. In Palestine, for example, daily practices of getting from here to there are a physical and psychological trial that entails enduring myriad checkpoints and often temporary and erratically placed road closures.[16] Moreover, the spatial regimes of control enacted by the Israeli government in occupied Palestine in the name of security are also a feature of state power elsewhere in the region and across the globe. In this book, I explore how, in an increasingly militarized post-2004 Beirut, moving through public space has entailed encounters with constellations of security that discipline and surveil, encounters experienced especially by members of disenfranchised populations that are understood as "other" or criminal. The sudden and variable character of these security formations requires residents to find ways to manage their daily course of activity in response to the appearance or news of a roadblock, checkpoint, or military installation set up in the wake of but also in anticipation of what is deemed a security threat.

Moreover, as sociologist Jack Katz (1999) has shown, experiences of mobility are fleeting, quotidian dramas laden with raw, emotional power.[17] When people move through Beirut, they are experiencing the intersection of citizen and state, of the more and the less privileged, and, in general, the city's politically polarized geography. The ability to pass through and around security blockades, for instance, often has to do with an individual's class presentation, which emerges from visible markers of class such as clothing, hairstyle, grooming, and means of transportation, along with behavioral disposition and accent or language.

For these reasons, I found that talk about mobility was an important way for Beirut's residents to convey sentiments about civic interaction and public life. When, for example, Beirutis inveighed against high-status show-offs who drove *mithl ma bidhun* (however they pleased) with no regard for others on the road or when they cited the state's lack of enforcement of traffic rules as a threat to public safety, they registered concerns about the workings of the civic order that were not only manifest in the context of traffic but that undergirded society more generally. The stories they told about getting around Beirut also positioned people as members of an urban community in a city marked by both present-day violence and that of the remembered past. The way that Beirutis engaged matters of mobility to talk about society, the nation, and politics therefore tells us something about urban life in a site of conflict and how relations of social inequality are engendered through spatial movement.

This book is about how the everyday spatial mobility that all people experience shapes and is shaped by local hierarchies of class and status. In my account, I show how ways of moving through the physical space of the city, from standing in the streets and hailing a ride in a shared taxi to navigating various kinds of security barriers as a pedestrian, are streetside interactions between the various kinds of people caught up, as anthropologist Lila Abu-Lughod would put it, "in [the] intersecting and conflicting structures of power" that span Beirut (1990, 42). In this way, my ethnography builds on studies of mobility that use the movement of people in urban spaces as a site of inquiry into power relations and is a departure from anthropology's long-standing focus on human movement across transnational borders. When anthropologists look at running urban errands by car (Jain 2002) or at motorbikes as unexpected agents of globalization (Truitt 2008) or at the affective experiences of taxi drivers and day trippers (Notar 2012a) or at how public transit is a mechanism of racial and class segregation (Czeglédy 2004), they are also looking into the implications of automobility for social life. Building on geographical, sociological, and historical studies of automobility,[18] these ethnographic studies of spatial movement have shown us how mobility practices and narratives order social stratification and frame civic possibilities.

Feminist thought has also transformed our understandings of space and place in similar ways. Feminist scholars have described how social and economic patterns of gender inequality are expressed in the organization of the cultural landscape at various scales—from domestic architecture to the boundaries between urban communities—and how these spatial patterns not only reflect but also reinforce differences of gender, class, and race.[19] In Middle Eastern cities, their research has focused primarily on how gender is used to divide and define public and private spheres and on the ways in which women access and move through gendered spaces by using, for example, practices of veiling and consumption.[20] The domain of gender is underexplored in urban studies of Beirut and clearly deserves a separate study. While issues of gender are treated in the book, I envision undertaking a future project about masculinity and public space that will allow me to expand on my investigation of the militarization and securitization of the city and build on emerging scholarship of masculinity in the Middle East.[21]

Citizens and the State in Lebanon

In the book, I use the term *urban citizenship*, by which I refer to belonging to an urban community, a belonging that I think of as being rooted more in a shared recognition of a city's ways of life and culture rather than in rights and responsibilities. I therefore conceive of the city as a locus of citizenship in the sense that urban residents feel a sense of membership in and identification with cities as a polity apart from—though often overlapping with—that of the nation-state.[22] In my ethnography, I highlight how classed and politicized notions of belonging to an urban community are formed through mobility experiences. In so doing, I aim to bring into focus the "tumult of citizenship" that anthropologists James Holston and Arjun Appadurai (1996, 188) describe. In this formulation the city constitutes neither the background nor the foreground for struggles among different groups but, as Reem's words previously in this chapter suggest, the battleground itself, through which individuals and groups define their identity, stake their claims, wage their battles, and articulate rights, obligations, and principles (Isin 2002, 283–284).

Social and cultural histories of Lebanon have understood citizenship largely through a focus on the growth of the country's sectarian-based political structure, which apportioned power along sectarian lines to groups of political leaders (*zu'ama*). This literature describes how the dominant class, comprised of a coalition of families from various religious sects, is buttressed by patron-client arrangements that require the loyalty and affiliation of Lebanon's citizens.[23] These kinds of political dynamics, and their intensification during and in the wake of Lebanon's war (1975–1990), have thereby made political sectarianism a kind of subnationalism that is primary in the making of Lebanese citizens. In my account, I draw on this understanding of Lebanese citizenship as being rooted in the country's political sectarian framework but also go beyond it by showing how class, status, and everyday relations with the state critically shape what it means to be an urban citizen of Beirut.

Spatializing the State

During summer 2013, in the context of a deteriorating domestic security situation because of the spillover from the Syrian civil war, billboards throughout Beirut aimed to send a comforting message about the state's interest in taking care of citizens. The billboards pictured five members of the national police agency, the Internal Security Forces (ISF). The billboard's message was "May you be well and in good health every year," an Arabic sentiment expressed on holidays and special occasions and, below, "Making the nation and the citizens content." With its kind message of well-being and smiling visages of police officers, the billboard cannily personalized and made accessible the state's security apparatus by foregrounding its workforce. This was a strikingly different political sensibility than those expressed by the banners, posters, and images that were draped throughout

FIGURE I.1. Internal Security Forces billboard: "Making the nation and the citizens content." (Photo by author.)

the city endorsing political figures and parties in a way that makes a kind of "state for all" secondary to party politics. As citizens moved through Beirut, the billboard seemed to remind them that the state was working to protect and serve ordinary people.

Yet the idea conveyed by the billboard, of a competent state concerned with the protection of all citizens, was one very much in question. Although mechanisms of state security were visible on the streets, people I spoke with during the time of my research generally associated these with the protection of VIPs rather than ordinary citizens. And, in the context of the so-called chaos on the roads, the state was often described as failing to ensure the safety of citizens by allowing dangerous mobility practices to go unchecked. In short, the prevailing understanding of the state I encountered was one that emphasized its shortcomings.

To build on studies of the state that aim to break down notions of the state as a coherent and rational monolith that sits "above society,"[24] I am looking here at the spatial dimensions of state power that appear in everyday scenes of getting around the city streets. In so doing, I move beyond the interpretations, prevalent in political science and policy studies, of the Lebanese state as weak or failed,[25] by showing that instead of being a unitary entity that is categorically weak or strong, state power is contingent on various kinds of social relations that are worked out, quite literally, on the ground, in the public fray between citizens and representatives of the state.

As a point of regional comparison, the character of the Lebanese state was also debated and discussed by the news media amid the uprisings in the Arab world in spring 2011. While Lebanon's fractious politics often make headlines, in this case it was the country's absence of political activity that drew attention. Why have there been no sweeping changes in Lebanon? the press asked. As Deeb and Harb write, "Not only is there no dictator against whom to revolt; divisions among Lebanon's political communities have been so deeply established over the course of the nation-state's modern history so as to preclude the stuff of united televised protests, most of the time" (2013, 33). Part of the story this book tells, however, is about the sense of frustration, disaffection, and disenfranchisement that emerge from Lebanon's flawed democratic, multiparty system, in which power is shared among a small group of leaders and families who claim affiliation with particular sectarian and ideological communities.

STRUCTURE OF THE BOOK

In chapter 1, I provide a historically informed overview of Beirut's built, physical, and transport environment that reveals the city's unplanned, informal, and privatized character and sets the stage for my ethnography of mobility through the city. Chapter 2 presents a picture of how historical experiences of war and conflict have shaped the city's space. I turn next, in chapter 3, to an ethnographic exploration of how this history, in relation to contemporary political sectarian violence and tensions, shapes residents' understandings of and movement through the city.

In the next three chapters, I show how people spatially negotiate the city in relation to social class, politics, and state power. Chapter 4 takes up the ways in which security practices and installations seek to enact a kind of spatial order in the city. This order—adopted mainly by and for an intersecting group of political and class elites—creates conditions of insecurity in the daily lives of ordinary residents. While chapter 4 focuses on the establishment of order in the streets of the city, chapter 5 focuses on how disorderly traffic is understood through an ethnographic investigation of talk about the "chaos" of getting around the city and about how, on the one hand, talk of a chaotic mobility unites different kinds of Lebanese under a political imaginary about citizens coping with a "developing" nation and corrupt government, while, on the other hand, hierarchical distinctions among citizen-drivers are engendered by this same discourse. In chapter 6, I focus on a particular type of encounter with political and state power in public space: interactions between traffic police and drivers. By looking closely at the work of the traffic police as a site of everyday state formation, I consider how the state is constituted and understood from the perspective of both citizens and street-level bureaucrats, who, despite, but also as a result of, their discord, share common ground in their sense of insecurity and disappointment with the state; this perception is conveyed through the oft-used expression *ma fi dowla* (there is no state). The Conclusion summarizes the key findings of the book by emphasizing the role of mobility in the public enactment of class, political, and state power.

1 • THE PRIVATIZED CITY

Forming a cape that extends into the sea, Beirut is situated at the geographical center of the Lebanese coast. Initially developed around its port area, the city expanded through the twentieth century in all directions: northward along the coastal plains and their neighboring mountainous areas toward the country's second largest city of Tripoli, eastward in the direction of the Beirut-Damascus highway, which rises into the mountains, and to the south, where, by the early 1950s, villages were being transformed into suburbs by rural migrants seeking economic opportunity in the capital and, later, several Palestinian refugee camps were established.[1] Today, about half of Lebanon's total population—close to two million people[2]—lives in the Greater Beirut area, which comprises the city and its suburbs.[3] It is a densely built landscape of concrete structures, "mostly indiscriminate mid-rise and high-rise buildings that cast their shadows on the remaining vestiges of villas or low-rise houses of the French Mandate period" (Verdeil 2005).[4] Buildings sit in close proximity to one another with narrow setbacks from the street; commercial shops and services are on their ground floors.

In many neighborhoods, towers of otherwise indistinctive apartment blocks express the individuality of their occupants through a particular architectural feature: the balcony. Beirutis speak about a kind of *ilfeh* (familiarity or intimacy) that characterizes the city's urban culture. The apartment balcony, situated on the border between the public space of the street and the private realm of the domestic sphere, is a site where this intimacy is cultivated. From balconies, which are used year-round, neighbors converse with one another, families' socks and undergarments are hung

FIGURE 1.1. Map of Beirut in Lebanon and the region. (Map by Richard Gilbreath. Adapted from map by Andrew Alfred-Duggan, ITMB Publishing.)

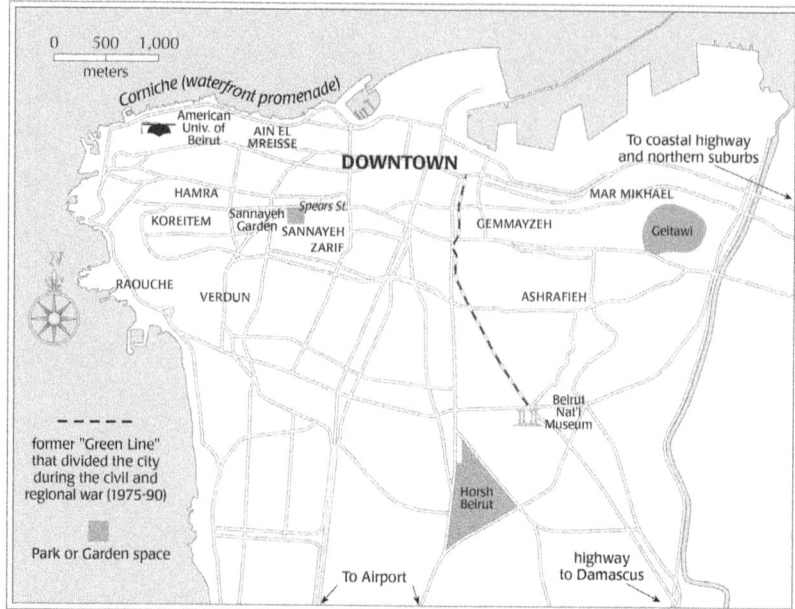

FIGURE 1.2. Neighborhood map of Beirut with main sites referred to in this book. (Map by Richard Gilbreath. Adapted from map by Andrew Alfred-Duggan, ITMB Publishing.)

out on clotheslines to dry in view of passersby, and residents shout down requests for grocery items to be carried upstairs by workers in shops below. Walking through the city or sitting at the open window of a public bus as it crawls through traffic, the streetside observer encounters these household scenes. In this way, the balconies help to produce an urban street culture that is caught up in the sights, sounds, and even smells of everyday domestic life, from the yells of children calling down to their friends in the street and the aromas of lunchtime meals wafting out from kitchens to the sounds of the television news and glimpses of residents tending to their hanging gardens, oases of green amid a dense city colored concrete gray.

Moving through Beirut, one also observes its "half-commercial, half-industrial" character (Adnan 1982, 9). The sight of tower cranes putting up high-end residences in and adjacent to the city's historic and geographic core, the downtown area,[5] gives evidence of the fact that the commercial traffic in land is one of the most important sources of private wealth in Lebanon.[6] In the east, just past the picturesque neighborhood of Ashrafieh and its late nineteenth- and early twentieth-century French-style architecture is Bourj Hammoud, an area that was founded by survivors of the Armenian Genocide and that expanded mostly during the 1930s,[7] where scenes of manufacturing and light industry—from furniture and shoes to mechanical parts and mattresses—emerge. From there, one approaches the junction for the coastal highway heading north at Karantina, where the air is thick with the putrid smells emanating from both the city's waste-treatment plant and one of the country's largest slaughterhouses. The coastal highway, one of the most congested traffic corridors in the country, is used daily by commuters living in the northern suburbs. These suburbs rise upward from the coastal plain to the storied Lebanese mountainside.[8] Moving north and east from the beaches of Beirut, through congestion and industry, to the foot of the Mount Lebanon range is thus a route across the city's diverse physical landscape, one that brings to mind a favorite aphorism often repeated to foreign visitors: "Only in Lebanon can you go to the beach and ski in the same day." To travel along this route in Beirut is also to move across different territories, areas of the city that are identified with particular sectarian and political groups that have been and continue to be adversaries.

In this chapter, I approach Beirut's physical landscape as a code whose deciphering may be undertaken through the study of its "ordinary but diagnostic features" (Meinig 1979, 6). Two processes, modes of privatization nourished by a laissez-faire market-led model of urban development and political sectarian conflict, have been the key power geometries (Massey 1994) shaping the city's space in the modern era. I consider these power geometries—by which I refer to the ways in which spatial arrangements, access, and mobility reflect hierarchies of power and control—in this and the following chapter. Here, I provide a historically informed overview of Beirut's built, physical, and transport environment that reveals the city's unplanned, informal, and privatized character.

THE RISE OF MODERN BEIRUT

Lebanon became a province of the Ottoman Empire in 1516, but Beirut was an insignificant port town for much of the Empire's rule. It was not until the late Ottoman period (1860–1914), when the city was made an imperial administrative center and trading activity shifted from the interior to the coastal region along the eastern Mediterranean, that Beirut came to prominence. Along with the emergence of the French-supported silk industry in the Mount Lebanon region, which expanded the export sector and stimulated ancillary enterprises in finance, shipping, banking, and insurance,[9] the city also developed as an outcome of administrative reforms instituted by the Ottomans to "modernize" the Empire; these reforms encouraged British and French investment in infrastructural projects in and around the city.[10]

These investments increased the economy and attracted rural-to-urban migrants seeking not only a better livelihood but also refuge from religious violence in their villages.[11] In the course of the nineteenth century, rural migrants transformed Beirut from a small town of six thousand people spanning a quarter of a square mile into a major seaport city with a population of one hundred twenty thousand by the century's end (Fawaz 1983, 1). While many of the these new residents in the city retained ties to the villages they left—and the building of roads that extended across the mountains and faster carriage service facilitated these connections—rural migrants to the city were, by the late nineteenth and early twentieth centuries, becoming Beirutis. They became Beirutis in a civic sense, as residents of an urban polity governed by a municipal council rather than through the charitable institutions, sectarian communities, and private property owners that collaboratively supervised villages (Abdou-Hodeib 2011, 478). But, in another sense, these rural-to-urban migrants, many of whom entered the trading milieu and constituted part of the city's burgeoning middle class of merchants and salaried professionals, were also becoming Beiruti through their participation in new kinds of public and social practices.

During this time, the city became both "the project and object of the cosmopolitan desires of an Ottoman-Arab bourgeoisie to belong to a distinctly modern epoch" (Hanssen 2005, 14). The interest of this urban middle class in taking leisure in the public realm gave rise to cafes, theaters, balls, evening dances, clubs, public gardens, horse racing viewed from within a European-style hippodrome, and roads designed specifically for owners of automobiles.[12] In the park squares, coffeehouses, art galleries, and theaters of turn-of-the century Beirut, new forms of public sociality took shape that brought the people of Beirut together both by chance as well as along class-based lines. With the appearance of these kinds of places, the removal of its medieval city walls, and its newfound status as an imperial capital, Beirut changed profoundly during the late Ottoman period; a modern city was inaugurated, one characterized by a vibrant, middle-class public sphere.

FIGURE 1.3. Cafe at a Beirut public garden during the late Ottoman period, ca. 1900–1920. (Courtesy of the Library of Congress, LC-DIG-matpc-01186.)

Allied victory in World War I brought the dissolution of the Ottoman Empire.[13] Following the declaration of independence in 1920 by an Arab Congress and its provisional recognition by the Allied Powers, France invaded Syria and, working together with the British, divided the Arab Near East of the Ottoman Empire into a number of separate states subject to colonial control. While the British Mandate administered Palestine (modern Israel, the West Bank, and the Gaza Strip), French mandatory control was established over six states that divided Lebanon and Syria.

The decision to create a State of Greater Lebanon—whose borders are those of the country today—which was separated from the states of Syria, had several important consequences. First, it further strengthened France's historic alliance with Lebanon's Christian communities, especially Maronite Catholics,[14] and expanded the political influence of these groups.[15] Second, the choice to create a Greater Lebanon in particular and French mandatory policy more generally confirmed the financial and commercial hegemony of Beirut over the mountain regions and the development of a pattern of economic activity in which agriculture and industry became ever more subordinate to banking and trade.[16] The legacy of this center-periphery mode of development is evident today in the considerable infrastructural, health, educational, and income disparities between the Beirut and Mount Lebanon regions and the south, Bekaa Valley, and the north.[17] Some have even gone so far as to argue that Beirut not only is the capital of Lebanon but, given its role as the country's economic and demographic center, has come to constitute a city-state.[18] Finally, the establishment of a French-controlled Greater Lebanon entrenched political sectarianism, an important development that I discuss further in the following chapter. French officials ruled through paternalistic power that distributed—and rescinded—benefits to the "ruled" through a mediating elite.[19] Although this elite brought Sunni, Shi'a, Druze, and Greek Orthodox leaders into its fold, it reinforced the notion of a sectarian-based political order.[20] As historian Philip Mansel (2010, 300) observes, even municipal appointments in the government of Beirut were apportioned according to sect, a practice that entrenched a sense of difference among residents of the city rather than integration.

As in France's other colonial territories, the French administration in Lebanon was a technocratic one that emphasized vast infrastructural projects such as the laying out of a cross-country road network, the modernization of the ports and postal service, and the creation of a telephone network, hospitals, and sanitary services (Picard 1996, 38). In Beirut, public space was made French through commemorative practices that named streets and squares for French military figures and that erected monuments and statues in the honor of French government figures and military men and also through the design of the built environment. The city's downtown center was a particular focal point for the mandate authorities, as it was intended to be the showpiece of French urban planning in the Levant. There, Haussmannian Paris was re-created.[21] A symmetrical pattern of long, straight, and wide avenues took the place of the Ottoman-era labyrinthine network of narrow alleys, open-air souks (markets), and crowded quarters. Streets radiated out, in a star-shaped design, from Place de l'Etoile (Square of the Star).[22] The construction of public buildings and residential structures also gave other parts of Beirut—like the Ashrafieh neighborhood on the east side of the city, for example—a distinctly French character.

Through its colonial cultural policy, the *mission civilisatrice* (civilizing mission), the French made Beirut the center of French culture and language in Lebanon. Secular and religious schools in Beirut—including Muslim ones—many of which had been opened along with American and British missionary schools during the nineteenth century, served as an ideal staging ground for this enterprise.[23] Efforts to cultivate French language and culture in Beirut, those that began well before the mandate era but intensified during the colonial period, have endured until the present. In today's Beirut, not only are aspects of French culture, foods, and fashion, for example, commonplace, but also French is widely spoken and the Lebanese Arabic used by Beirutis is peppered with French words and phrases.

In the late Ottoman and French colonial periods, Beirut developed from a backwater imperial holding into a flourishing trading center with a lively middle-class public sphere. In the following sections I explore how modern Beirut has taken on an unplanned physical form characterized by state-supported processes of private investment, shrinking public space, and vehicular congestion.

LACK OF PLANNING AND INFORMALITY

Beirut's Ottoman and French history are both evident in the city's built environment today, as part of an eclectic architectural mix that has late nineteenth- and early twentieth-century European styles mingling with 1960s modernism and the faux-tradition of the city's recently redeveloped downtown.[24] The diverse architectural styles provide evidence of the periods of the city's growth and are but one feature of the patchwork character of the urban landscape as peddlers pushing carts loaded with seasonal produce share the street with global conglomerates like Starbucks and H&M and newly constructed high-rise luxury buildings rise up on narrow blocks, dwarfing their two- and three-story neighbors. "In Beirut, it's anything goes," one architect told me; "the city has just sprung up in every direction, there is no plan." Although its development has been the object of careful and comprehensive state-sponsored planning since the mid-twentieth century,[25] these plans for Beirut were never realized, and, as a result, the city's physical texture is a pieced-together one.

The urban planners, architects, and engineers I spoke with in 2004–2005 about the organization and design of a city that is often described as haphazard, unplanned, and unregulated[26] drew links between the lack of a comprehensive vision for the city and the political infighting for which Lebanese state institutions are well known.[27] The massive development project under way in the city's center, undertaken by the Solidere corporation,[28] was cited by several planners and architects as an example of how the private sector has been a more efficient player than the state in planning and improving urban space. During our conversation in spring 2005, architect Robert Saliba, for instance, described how the "corporate approach to urban design has proven to be more effective in Beirut in comparison with the ineffectiveness of the traditional, governmental approach." This privatization of urban planning and development, a process that in the Arab world has seen the state selling off publicly held lands to the highest bidder with little regard for the public interest,[29] is understood by some, in the context of Lebanon's fractious politics, to be a redemptive one. For example, at a public lecture in May 2005, Amira Solh, senior urban planner at Solidere, spoke about how having urban planning in the hands of the private sector is "safer" in Lebanon because sectarian and political tensions always take over in the public realm and then "everything devolves into these [political] conversations, debates and disputes and becomes stalled."[30]

Like Solh, architect Nabil Gholam described the private developer as being free from the polarized politics that undergird municipal governance in Beirut. "The municipality's plans for the city," he said during an interview in November 2004, "are not planned at all; they're spontaneous. Even single owners can change the city's space. . . . Just down the street from here, there's a small company and they just took a strip of land in front of their building and completely changed the sidewalk!"

This lack of planning and standardization is also linked with the problems in the traffic infrastructure. In a summer 2010 meeting with planners at Majal, an urban planning institute, one staff member explained to me that the municipal government is well known for its complete lack of planning: "The municipality comes to us for technical assistance, but they always come when it is too late. Once an area of the city has a problem—like now with Gemmayzeh, they are aware that it is too crowded, no service [taxis] will go through there, no buses are able to get through—that is when they come and try to find a solution. But there is no planning ahead." Similarly, in our conversation in spring 2005, transportation engineer Youssef Fawaz described how there is no set pattern of design or implementation for infrastructural elements like speed bumps and parking bollards. He continued, "These things should be standard, but there's no big picture. There are no specs [specifications]; there are all different sizes, shapes, and colors. I know a group of residents who were just able to install their own bollards on their street; . . . the municipality agreed that they can make them whatever shape and size they want!"

Drivers and pedestrians experience this lack of planning and standardization in their everyday mobility. The fact that setbacks are not regulated (building owners can build any distance away from the main road and from the edge of the sidewalk) produces inconsistent streetscapes.[31] Curb and manhole-cover heights are variable and so, to avoid stumbling while on foot or damaging a vehicle while driving, a certain vigilance when getting around is required. Cars block sidewalks and force pedestrians to walk in the street, and landmarks are used to navigate the city and to give directions because of the lack of standardized names for streets, the absence of numbered street addresses, and the inadequate signage.[32] Another example is a service road that runs parallel to and offers an alternative to travel along the often-congested coastal highway. Traffic changes direction during different times of day along this road, but these times are neither posted nor standardized. In short, drivers cannot plan ahead to take the alternative road but instead must cross over to it to find out which way traffic is flowing.

This kind of unpredictability is also a feature of public transit. In neighborhoods with narrow one-way roads, drivers behind buses become incensed, laying on their horns, waving their hands, and leaning out of their car windows to shout for the buses to move. The fact that there are no fixed bus stops—an inherent aspect of the bus's informality—is a benefit to the bus passenger, who can board and disembark anywhere along the bus route rather than worry about having to reach and wait at a "formal" and marked bus stop. But, for other vehicles on the road driving behind the bus, the lack of fixed stops—and the absence of lanes where buses could move out from the flow of traffic—is a nuisance.

Various other kinds of ad hoc practices constitute the city's physical, transport, and built environment. As part of the process of informalizing state structures in the contemporary era, urban development in cities across the globe has increasingly moved away from state legislation and public deliberation toward behind-closed-doors agreements arbitrated between state and private actors.[33] In Beirut, the expansion of informal decision making and the allowance for exceptions to the law with regard to the built environment since the turn of the new century have brought about a relaxation of the state's regulations regarding maximum building height, construction permits, and the acquisition of property by non-Lebanese.[34]

This state-supported process of planning by informality has freed up more and more of the city's space for high-end—and high-rise—real estate aimed primarily at buyers from the Gulf Arab and Lebanese expatriate communities.[35] These are short-term, typically summer, residents; Rahif Fayad, an architect and cultural critic, described them in our interview in spring 2006 as having "no stake in our society." Housing production is not only an arena for profit making, however, but also an opportunity to expand and affirm political constituencies. In the wake of the summer 2006 war with Israel, for instance, Hizbullah's rebuilding of the Haret-Hreik neighborhood in the southern suburbs vis-à-vis its newly established private development agency, Wa'ad, demonstrated the central role that nonstate actors play in shaping the physical and political geography of Beirut, particularly at a time when national sovereignties and political identities are increasingly contested.[36]

Informal settlements that developed along the city's periphery beginning in the 1940s are another feature of Beirut's landscape. Built in violation of urban and building regulations (Fawaz 2009b), these settlements became home to rural migrants from poor (predominantly Shi'i Muslim) parts of the country who acquired land, built housing, and accessed services in less desirable and undeveloped areas of the city. Residents in these areas, however, now find themselves increasingly displaced because of increased land prices and the territorial expansion of urban-development projects. Informal housing and infrastructure are also in plain view at some of the city's construction sites where mainly Syrian laborers are employed. Some of these laborers live where they work, setting up semi-permanent encampments just beyond the construction façade in parts of unfinished structures. Signs of residence, like laundry hanging over the construction walls to dry, are a common sight at these incomplete buildings. The tangles of power wires that haphazardly drape over street blocks throughout Beirut, to give another example, are suggestive of residents' practices of informal energy hawking by hooking into power lines and by meter pirating.[37]

In this way, a range of actors and processes come together to produce Beirut's informal urbanization. There are the private power holders who informalize modes of urban development by using their influence to bend state regulations to their will—for example, by bringing about legislative changes regarding maximum building height and construction-permit procedures. There are also the ordinary and underserved users of the city who establish housing and hook into infrastructural services through illegal and informal means. And, through its rebuilding of the Haret-Hreik neighborhood, Hizbullah has employed informal channels, such as word-of-mouth communication to residents about reconstruction procedures and options (Fawaz 2009a, 326–327), to enhance its authority as a private urban planner and developer, thereby consolidating its territorial base in and claims to parts of the city in the process.

AUTOMOBILITY AND PUBLIC SPACE

The hegemony of the private automobile is another dimension of privatization in Beirut's built and physical landscape. Beirut's public tramway lines, which first appeared in the early twentieth century, were dismantled in the mid-1960s to allow the ever-increasing number of automobiles greater freedom of movement. The disappearance of the tramway in the 1960s was followed, later in the 1980s, by the demise of rail transport, which had aided in the shipping of freight from the port of Beirut.[38] These changes in the transport scene only intensified vehicular congestion in a city that had already been described, in 1963, as having "a very acute traffic problem caused by the increase in the number of cars and the small number of new streets" (Riachi 1963, 111). Once championed as a boon to the country's tourism sector,[39] cars were, by the 1960s, thought to detract from the pleasures of visiting Beirut so much that the special Tourist Brigade of traffic policemen charged with fining drivers for the misuse of their horns was formed.[40] This was a time many consider to be Beirut's golden age, when the city attained the moniker "Paris of the Middle East." Beirut became in the 1960s not only a fashionable destination for European and American jetsetters and a city of pleasure for those seeking the sun and nightlife but also the literary, publishing, and entertainment capital of the Arab world. As the end point of the Trans Arabian oil pipeline, the city had begun to benefit from its links with the newly oil-rich Gulf. But it was also an increasingly congested city, one whose population doubled between 1955 and 1965. By 1973, the bus played only a minor role in getting people around, as most daily journeys in the city were completed in passenger cars.[41] By the 1970s, then, Beirut had developed an urban culture dependent on cars.

When I spoke with people about the roots of the city's traffic problem, they often commented, "there are too many cars" and "there is not enough good public transportation." The Ministry of Public Works and Transport estimates that about three million daily motorized trips occur in the Greater Beirut area with trip mode split among private cars (68 percent), shared service taxis and private taxis (15 percent), minibuses (11 percent), and buses (6 percent).[42] Complaints I heard about buses in particular always had to do with their low frequencies, slow and variable and unreliable travel times, and poor geographic coverage, as a traffic-policy report also found (Aoun et al. 2013, 53). Maya, a young woman in her early twenties, told me that when she took the bus from near her home in the northern suburbs to her high school in Ashrafieh, she did not mind that it took so long. "I would just bring my books with me and study the whole way, so it wasn't so bad. But it was so slow, I can see why no one wants to take the bus!"

Congestion not only is produced by Beirutis' reliance on passenger cars but is also an outcome of the city's narrow roads and lack of underground parking. In summer 2013, in a sign that the traffic had only worsened since my last stage of research in 2010,[43] service drivers often refused to take me as a passenger when my destination was in an area where they anticipated heavy traffic. Or, they would ask me to pay *service-ayn*—a double fare.[44] To get where I needed to go, I often told service drivers a destination that was close to where I was going but on the periphery of a very high-density corridor—*awal Hamra* (at the beginning of the Hamra neighborhood) instead of just Hamra, for instance—so that the driver would agree to take me. The city's vertical expansion, through the development of high-rise towers, has only exacerbated the problem by increasing the number of city residents (with cars) per unit area of land and reducing the number of vacant lots that can be used for parking.[45] Although a new law mandates that new buildings provide underground parking for their residents, it is doubtful, as architect and urban planner Gregoire Serof conveyed in our interview in 2010, that this law is being strictly enforced. Congestion caused by irregular modes of on-street parking—double and even triple parking are common—further reduce road space and hinder vehicular movement. According to the Ministry of the Environment, in 2001 the average vehicle speed in Greater Beirut was around 20 kph (12 mph), and free-flow travel time was typically doubled or tripled because of delays (Aoun et al. 2013, 53).

The traffic of private cars not only contributes to the feeling, as one respondent put it, "that there is no place to breathe" but is also a key feature of Beirut's privatized urbanism. Together with the takeover of public space by cars and state disinvestment in public transportation, there is a lack of free and open public spaces—playgrounds, sanitary public beaches, designated sports areas—that enhances the everyday experience of congestion in the city. Residents use the Corniche, a five-kilometer seaside boardwalk, like a park. In the cooler morning hours, when the air is clearer before the onset of traffic, it is a pathway for fitness walkers and joggers. On warm evenings, the Corniche becomes crowded with leisurely amblers of all generations, with others perching along the promenade with their argilehs (water tobacco pipes) and deck chairs in a circle. But parks in the sense of green areas that offer trees, distance from traffic, and safe play areas for children are scarce.

The largest green space, a lush landscape of pine trees known as the Horsh (forest), has never fully reopened to the public since its restoration after the protracted civil and regional war that ended in 1990. While a nascent movement to enhance existing gardens and green spaces led by NGOs and private foundations is pressuring the state to build new

FIGURE 1.4. The Corniche. (Photo by author.)

parks,[46] the city in fact looks like it perhaps will become less, rather than more, green in the near future: in June 2103, the municipal government announced plans to demolish—and later replace parts of—a park known as the Jesuit Garden in Geitawai, a neighborhood of winding, narrow lanes on the eastern side of the city, in order to build a parking garage. Vigorous protests against the project ensued and, as of this writing, have led to its delay as the municipality undertakes an environmental impact study it says will be reviewed before a decision is made about moving forward.

The state's emphasis on establishing more parking rather than parks and its leaving the creation of a more efficient bus system at the bottom of the list of priorities shape not just the urban landscape but also how people live. Air and noise as well as environmental degradation threaten residents' health and quality of life.[47] The cityscape's consumption by the car also threatens civic life, as a decline in parks and public spaces reduces opportunities for different groups of people to come into contact with one another. As was the case in the 1960s, when the tramway was eliminated, state agencies in Lebanon today continue to support transformations of Beirut's built and physical environment, like the construction of multilevel parking garages and new highways, to accommodate the private car.[48]

"THE GROWTH MACHINE"

"In The City, this center of all prostitutions, there is a lot of money and a lot of construction that will never be finished. Cement has mixed with the earth, and little by little has smothered most of the trees. If not all" (Adnan 1982, 9). Though written more than three decades ago, Etel Adnan's description of Beirut in her civil war novel, *Sitt Marie Rose*, is still apt, for the city exists in an enduring state of construction. New buildings go up everywhere, soaring taller and taller in the quest to offer their residents unblocked sea and mountain views. Some new buildings fill in the city's few vacant spaces while others take over lots where now-demolished older buildings and homes once stood. Tower cranes are a fixture of the city's skyline as processes of state-supported privatized and market-let urban development "grow the city" in pursuit of real estate investment and value.[49] Led by both private and public actors, the city's growth has materialized through unplanned and informal means and has produced an urban physical landscape in which the hegemony of the car has gone unchallenged. In continuing to choose highways and parking garages over the preservation of the city's historical social fabric, the planners of Beirut, in the words of Lebanese architect Hashim Sarkis (2014), "reveal an outdated understanding of the contemporary city," in which the municipal administration aligns itself with the developers in opposition to local citizens.[50] While the city's growth in high-end real estate and highways is almost always portrayed by power holders as beneficial to everyone, the uneven distribution of this growth has reinforced inequalities between central Beirut and its outlying areas and, moreover, has positioned affordable housing, public transportation, and free public spaces as low urban-development priorities for both the municipal and national governments. In these ways, the state has acted neither as a regulator of Beirut's physical landscape nor as a public benefactor, and it has diminished ordinary people's right to access the city in the process.[51]

As is clear from the discussion above, the physical space of modern Beirut has been significantly shaped by modes of privatization involving deregulated and informal building practices, the diminishment of public lands, and the dominance of the automobile. As one of the city's key power geometries for organizing spatial access and mobility, these modes of privatization reflect hierarchies of power and control and have created a city congested with traffic and beset by ongoing construction projects, a city in which residents say it is increasingly difficult to find space to breathe and room to move. Political sectarian conflict has also played a central role in shaping the way that Beirutis experience urban space. This conflict is the focus of the next chapter.

2 · THE SPACE OF WAR

At the National Museum of Beirut, built during the French Mandate, glass-encased objects lead visitors through a telling of the city's ancient geopolitics, one focused on a chronology of conquerors and empires. Beginning with the Bronze and Stone Ages, the museum charts the city's role as a maritime coastal trading post during Canaanite, Hellenistic, Phoenician, Roman, Persian, Byzantine, Arab, Crusader, and Mamluk eras, creating what art historian Heghnar Watenpaugh calls an "invented tradition of cosmopolitanism as a national trait" (2004, 200).[1] But on my visit to the museum, another more lively account of Beirut as a Mediterranean "crossroads of civilizations" was also visible. Upon entering, I encountered fifty or more schoolgirls in their early teens. Dressed in hijabs and olive green *'abayat*—an Islamic garment that covers the length of the body—the girls were excitedly hurrying around the place in groups of twos and threes. They would stop in front of a display case or figure for only a brief second before shuffling off to another down the corridor, chirping and giggling while en route. I asked two of them what school they attended, and they told me that it was near the airport in Dahiya, the predominantly Shi'i Muslim southern suburbs of Beirut. Heading up the stairs to the second floor, I passed a nun in a blue habit and children in school uniforms, ostensibly students from a Catholic school. Upstairs, I came upon another group: boys wearing t-shirts and sneakers with a type of baggy black pants (*shirwal*) and white cap (*tarboush*) that is the traditional attire of Druze males, a religious minority community in Lebanon.

While both depict a city vibrant with diversity, the antiquities told of an ancient "before" whereas the schoolchildren seemed to represent a modern "after." And yet neither text of the city made immediately available to the observer the painful story about a place that has been made vulnerable not only by conquests, defeats, and natural disasters but also by the coming apart of the diverse nation that Lebanese novelist and sociologist Halim Barakat once argued "was a mosaic composed of several different groups that lacked consensus on fundamental issues" (1973, 301).

When the violence that evolved into a period of protracted conflict broke out in 1975, the National Museum's narrative of Beirut was tragically upended; the city characterized as the crossroads of civilizations became a crossroads of war. Situated on the Green Line, which divided a mainly Christian East Beirut and a majority Muslim West Beirut during the fighting, the main road next to the museum became one of the few crossings where people could traverse from one side of the city to the other during times of truce. The Mathaf-Barbir Crossing, as it was known, stood at a strategic location along the front and thus became too a site of killings and kidnappings. Occupied by various militias as well as the Syrian and Israeli armies, who used it as a bunker, as barracks, and as a staging ground for shelling and sniper fire, the museum was badly damaged during the fighting.[2] In this way, the museum not only fell victim to the war but also became, like other public spaces in the city, its instrument.

Although it is often called a civil war by scholars and analysts, the war in Lebanon was in fact a multinational one fought by members of Lebanese, Palestinian, Syrian, Israeli, U.S., and other foreign militia groups and state armies.[3] Given its domestic and as well as regional dimensions, following Sami Hermez (2012), I use interchangeably the terms *Lebanon's war*, *the long war*, and the *civil and regional war* rather than *civil war* to refer to the prolonged period of conflict (1975–1990) that made the country a strategic battleground for both internal as well as regional power plays. During the war, Beirut was spatially divided along sectarian and party lines, and militias, who held divergent ideologies and claimed affiliation with particular sectarian communities, governed, and enforced the boundaries of, parts of the city. Today, more than twenty years after the end of Lebanon's war, the neologism *Beirutization* still serves as a metaphor for the dissolution of community and the territorialization of a space into warring parts. Inhabitants of this space, as the metaphor also conveys, are unable to move freely as their movement is caught up in a geography of closely guarded boundaries that mark off friendly and enemy territories. As shorthand, Beirutization and its proxy, Balkanization, are spatial renderings of complex histories of political conflict and violence.

In this chapter, I explore how the war and postwar reconstruction have shaped the city's built environment as well as residents' experiences and interpretations of Beirut's urban space. To explore the spatial politics of this conflict, I draw considerably on scholarly work that is specifically focused on the war in Lebanon and its memory. During my research, I did not generally interview people about their experiences during the war, although the subject often came up in the context of talking about the current political violence or their patterns of mobility through different parts of the city. In this sense, the war did not frame my questions, but it often emerged in people's responses.

THE CONTEXT OF WAR

Because of its multifaceted character, it is difficult to summarize the war as a series of precise and delimited causes, events, and results. What I will attempt to do here instead is to outline some of the regional and domestic political tensions that were developing in the years prior to the outbreak of armed conflict in Beirut in 1975. By providing this brief outline of the complex national and regional geopolitics during this historical moment, I seek to displace the notions of inevitability that are often used in describing how and why this war happened. At times, in talking about the war in Lebanon with people in the United States, I have heard versions of the sentiment "those people have always been fighting with each other" aired. This idea expresses something of the conception of Arabs and Muslims as violent, driven by primordial desires, and unquestioning when it comes to their political allegiances. A historicized perspective of the war challenges understandings of political sectarian conflict in Lebanon as both timeless and inevitable. Rather than a story about the timelessness of sectarian conflict in the Middle East, Lebanon's war was the outcome of a convergence of issues at a particular historical moment that triggered sectarian and political hostilities.

The 1926 constitution of the French Mandate of Lebanon declared it a republic rooted in individual rights and liberties and political and juridical equality as well as communal rights and representation among the various sects.⁴ At the time of national independence in 1943, an unwritten agreement called the National Pact, which was considered the political cornerstone of Lebanon, was declared. A compromise among the country's ruling elites, it stipulated that Lebanon was a sovereign state, both Western and Arab in character.⁵ It divided parliamentary seats according to the relative population of each sect based on the outdated 1932 census figures (the only official census in Lebanon's history), which counted Christians as 52 percent of the total population and Muslims as 48 percent.⁶ The principal political and administrative roles went to the six largest communities: Maronite, Greek Orthodox, Greek Catholic, Sunni, Shi'i, and Druze. The president and commander of the army would be Maronite, the prime minister Sunni, the head of the parliament Shi'i, and the vice president of the parliament Greek Orthodox.

During his presidency, George W. Bush often cited Lebanon to legitimize the U.S. invasion of Iraq. A flourishing democracy is possible in Iraq, he reasoned, because one exists elsewhere in the region of the Middle East, in Lebanon.⁷ What the National Pact reveals instead, however, is the limits of Lebanon's democracy. The balance achieved by the pact made no provision for making adaptations to the apportionment of power in accordance with demographic changes but rather represented a partnership between the president and the prime minister that was intended to increase the participation of Muslims in the power-sharing apparatus of the state in order to redress the legacies of Maronite, and generally Christian, dominance in the public service. It was a partnership that made no commitment to popular consensus and preserved "sectarian-ness" as part of the nascent nation's political foundation. In time, the power-sharing framework of the National Pact would become vulnerable to collapse in the face of growing domestic tensions that were inextricably caught up with regional political developments.

In 1958, the National Pact was challenged by a political crisis that escalated into a civil conflict claiming an estimated two thousand to four thousand lives (Picard 1996, 74). The conflict emerged from various domestic tensions including political patrons' dissatisfaction with the corrupt rule of President Camille Chamoun, popular demands for greater state-sponsored social services like a social security system, and, most significant, the clash between two competing visions for Lebanon's identity. Post–World War II events in the Arab world, such as the Arab defeat in the war for Palestine and the founding of Israel in 1948, the rise of pan-Arab nationalism heralded by Nasser's leadership in Egypt, and the military coup in Syria in 1949, took place amid increasing U.S. intervention in the political affairs of the region as part of the Cold War foreign policy to hinder Soviet influence. These developments produced increasingly stark ideological battle lines in Lebanon between the conception of Lebanon as an Arab nation that should ally itself with the pan-Arab political movements in the region and the idea of Lebanon as being primarily affiliated with Western interests. With the establishment of the United Arab Republic (1958–1961), a union of Syria and Nasserist Egypt, the pan-Arabists in Lebanon began to demand that their country join the union while the majority of Maronites and other Christian communities called for Lebanon to remain aligned with Western powers. A military coup of the pro-Western Iraqi monarch led President Chamoun to call for U.S. assistance in defeating the pan-Arab opposition. More than ten thousand U.S. Marines and Army paratroopers were deployed in Lebanon—famously landing on the beaches of Beirut—and the forces in opposition to the government were suppressed. Although this moment of civil conflict and foreign intervention in regulating domestic tensions was brief, it was a harbinger of things to come.[8]

Although the 1960s are remembered as the good times, this was also a period during which economic inequalities intensified. The oligarchy that ruled both the political and the economic systems was characterized by class solidarity across sectarian lines. The poor were also a mixed (sectarian) group, although the Shi'a were the largest impoverished community and were concentrated mostly in the historically underdeveloped and mainly agricultural southern and eastern Bekaa Valley regions. With little to no government efforts to redistribute wealth or protect wages and no national policy for agricultural subsidization, an increasing number of Lebanese left their agricultural villages for Beirut in search of better opportunities. In other words, the blessings of this "golden era" were bestowed neither on all Lebanese nor on all Beirutis. Structural inequalities created by the country's service-oriented economy created an underemployed and disaffected urban population. Between 1967 and 1975 the cost of living had doubled, and, on the eve of war's outbreak in 1975, the unemployment rate was 15 percent, and 5 percent of the population controlled 50 percent of the GNP.[9] In this sense, as anthropologist Suad Joseph observes (1983), the coming crisis was a form of political and economic opportunity for the disenfranchised and displaced inhabitants of Beirut.[10]

During the 1948 Arab-Israeli war and following the establishment of the state of Israel, nearly one hundred thousand Palestinians fled to Lebanon. Lebanon granted citizenship to thirty thousand of them, most of them Christians.[11] Through the work of the United Nations Relief and Work Agency (UNRWA), most refugees were settled in fifteen camps on the edges of the cities of Beirut, Tripoli, Saida, and Tyre.[12] Twenty years later, after the Arab defeat in the 1967 war with Israel and Black September in 1970—an armed conflict between Palestinian militant groups and the Jordanian army—tens of thousands of Palestinians came to reside in Lebanon.[13]

Having been driven out of Jordan, the armed resistance group the Palestinian Liberation Organization (PLO) established its new base in Beirut. In the Lebanese refugee camps, the PLO and other Palestinian militant groups and organizations grew in strength, visibility, and number and began to clash regularly with Israelis along the border. One front of the war, then, had been opened before 1975, for, as historian Elizabeth Picard observes, "between June 1968 and June 1974, the Lebanese army counted more than 30,000 Israeli violations of their national territory" (1996, 83).

The myriad weaknesses of the Lebanese army were exposed during this period. Small and poorly equipped, the army appeared passive in the face of both devastating Israeli incursions and the activities of the Palestinian resistance groups that threatened the nation's security. The kinds of domestic cleavages and debates that had persisted since national independence, about the positioning of Lebanon vis-à-vis the West, on the one hand, and pan-Arab nationalism—which now supported the Palestinian armed struggle—on the other, intensified under these conditions. As a tiny country with a weak military, a sectarian political structure, and Syria and Israel as its neighbors, Lebanon appeared unable to prevent itself from being destabilized.

The specific event that is said to be the catalyst for the war took place in the southern suburbs of Beirut, an area where both Christians and Muslims lived. There, a shootout occurred between members of the Lebanese Phalange (a Maronite Christian–affiliated party), a supposedly Palestinian armed commando, and a bus carrying about thirty Palestinians who crossed the city suburbs to travel between refugee camps. While even today there is no commonly agreed on version of this event and its details remain contested, it is clear that this event set off a chain of fighting that evolved into the long war. Once the Lebanese army came to be dominated by the Phalange and fought against the Palestinian resistance fighters in early 1975, more and more Lebanese began to take either the side of the Phalange or that of the Palestinian resistance and to form factions of their own. In this way, the conflict expanded outside of the realm of Palestinian versus Maronite. In her novel *Sitt Marie Rose*, Adnan describes the scene in 1975 Beirut: "Many factions participate in the general terrorism, but the principal protagonists remain the Christian right and the Palestinian refugee-militants, that the former seems to want to eliminate entirely. Very rapidly the combat takes on the aspect of a civil war, and one that will last. The April air is perfumed and warmth mixes with freshness. The artillery booms. The local militia has even greater fire power than the regular army" (1982, 12). The ensuing war had many phases and fronts,[14] but in the following section, rather than provide a detailed summary of its many stages, I offer a sense of how the war transformed civic, spatial, and daily life in Beirut and its significance for the city's residents.

THE TRANSFORMATION OF CIVIC, SPATIAL, AND DAILY LIFE

During the war, political parties and organizations developed into militias and Beirut's neighborhoods became recruiting grounds. During my research, I met a man who told me that he had fought in the war. He looked to be about my age or younger, so I began doing the simple math to determine how old he might have been at that time. Sensing my thought process, he volunteered that he had been about thirteen when he first joined the Communist Party's militia.[15] "I had a friend," he said, "a guy from my neighborhood; he was a few years older than me, someone I looked up to. He was killed, and that's when I decided that I had to fight." He added that he was Shi'i, but that the Communist Party's militia had soldiers from all different sectarian backgrounds. His comment was a reminder that sectarian background and political affiliation were not one and the same during the war or before or after the conflict. In other words, to essentialize sectarian identity in Lebanon, to assume that this category wholly constitutes and defines individuals and communities, leaves one with an incomplete and often imprecise picture of the complexity of Lebanese politics and the political lives of ordinary people. The groups and militias involved in the war were numerous, and some, like the Communist Party, were founded on a politics that was, in part, pan- or antisectarian. It is also important to note that the fighting was not between the communities in the country, but between militias and armies. Some of these groups proclaimed that they represented particular communities, but this assertion in no way meant that all members of a community were in support of that militia. And, moreover, a great number of clashes during the war even took place between militias that identified with the same sectarian group.

As the nation became a contested terrain, host to soldiers both domestic and foreign engaged in a battle to determine to whom Lebanon would belong and through which rival regional powers it would be sponsored, Beirut's geography and that of the rest of the country were divided into distinct political zones. Each "canton,"[16] maintained its own network of foreign relations and enforced the borders of its territory. The most infamous of these borders was the Green Line of Damascus Road, which began downtown near the waterfront and extended southward dividing the city into two sides: a mostly Muslim west and a mostly Christian east. Lina, a university professor in her mid-thirties, grew up during the war. After graduating from college, she went to the United States to pursue a master's degree. She described to me how she lived in West Beirut, but regularly crossed the Green Line during times of truce to play and have sleepovers with friends who lived in Ashrafieh, a predominantly Christian neighborhood just on the eastern side of the dividing line. She recounted how the experience of crossing the Green Line was evoked during a graduate school orientation in New York City:

> During the war, I would cross over [to the east] at the *mathaf* [National Museum] crossing, and, to keep out of sight of snipers, we'd walk as close as possible to the buildings along the street, and we'd avoid being near other people; we'd be kind of moving along the walls of buildings. Then, in New York, at the orientation for international students, the police came to talk to us about how we should stay away from the walls and buildings along the street when we were walking alone. They said, "Keep away from walls, at night especially." This was exactly the opposite of what I did during the war! And then they

went on about the park near the school; they said, "Don't walk across that park at night, you'll never make it out alive. You won't reach the other side." I couldn't believe what I was hearing, and I was coming from Beirut!

As Lina describes, ordinary movement during the war required particular kinds of spatial and embodied practices in order to avoid danger in a territorialized city occupied by sniper positions. As militias took control of territories that ranged in size from sectors, which were no more than a block of buildings, to entire neighborhoods, the mobility of residents was circumscribed under their authority. Entry and exit points were patrolled by and subject to the will of armed militia members. These checkpoints became deadly during some moments of the war. If individuals could not prove to be persons permitted entry or could be identified—on the basis of the national identity card, for instance—as members of an "enemy" political sectarian group, they might be killed. Hence, "whatever the wealth and diversity of their backgrounds, social status, profession, class, ideological or political orientation, individuals came to be identified exclusively by membership in a particular sectarian group" (Picard 1996, 148).

During the protracted conflict, everyday life was subject to the episodic fighting that erupted among the militias. Dana, a young woman in her early twenties, recalled how the war disrupted her schooling. "For one year I didn't go to *lycée* [French secondary school] because the violence was so bad; my Mom homeschooled me. School would open and close all the time.... Every night we would just watch the news to find out if we were going to school the next day." Once, she said, students could not leave school—on whose grounds tanks were parked—because nearby fighting had intensified. "We just slept over there, in the sports stadium.... It's weird to say it now, but we were kids;... we thought it was fun. After the war was over, near the gymnasium, my friends and I found a whole room stored with Kalashnikovs." Militias had used the school to store weapons, she said.

Both the public and private built environments were shaped by daily fighting in Beirut and acquired new geopolitical meanings. Exterior walls of buildings became the protective cover for residents trying to stay out of the line of sight of snipers, and inside these buildings metal vault-like security doors were affixed over the doorways to the apartments. During the war's worst phases, life became entirely insular, as residents left the streets for the safety of the interior. On nights of bombing, families would sleep in their hallways to avoid proximity to windows. Or, they would join the other residents in the building for a night in the basement, which would serve as a bomb shelter. The war had a distinctive provincializing effect on social relations as it not only drove residents inside and limited the possibilities for civic life but also broke down trust and associations between people from different political and sectarian groups.

The war continued from 1975 until 1990. There were moments during which fighting would stop, and ordinary life would return—a school term without interruptions, for instance. And then this normality would be cut off again. Sustaining oneself through these many years of violence, the intervals of war and not-quite peace, an existence rife with unease and trepidation, required endurance and resolve. The stopping and starting of the hostilities necessitated just that, the ability to stop and then start again, the strength to rebuild, reopen, to reinforce optimism in the face of devastation. People who had the means to leave the country did, and wartime exiles flowed from Lebanon. There were those who had the kind of connections that furnished visas to the United States, Canada, Europe, and Australia, and there were others who left for African nations or the Arab Gulf. In some cases, men set up their families near relatives or other Lebanese in these locales and then returned themselves to Lebanon to keep businesses and shops going in spite of the war. Some of the wartime exiles returned to Lebanon later, in the 1990s and afterward, while others remained settled in their new countries.

Along with the Syrian incursion and engagement in the war that began in 1976, there was significant foreign participation in and contribution to the conflict. The Israeli invasion of 1982, and the massacres of thousands of Palestinians committed under their army's direction at Sabra and Shatila in southern Beirut in the same year, marked an exceptionally horrendous and horrific moment in the war. The Israeli occupation especially transformed the lives of Lebanese in the south. Many stayed, many resisted, and many left to settle in Beirut's southern suburbs. Moreover, the Israeli occupation, which lasted until 2000, ten years after the end of the war in Lebanon, brought increased support to Hizbullah and the organization's politics of resistance against Israel.

One of the militias that emerged during the war, Hizbullah, a Shi'i resistance movement with transnational links to Iran, was formed in the wake of the Israeli invasion and arrival of the U.S. military in 1982. Founded to defend Lebanon against the incursion of foreign occupiers, the group gained adherents and fighters from the largely Shi'i south and, later, from the southern suburbs of Beirut. And "while most militias either gave up their weapons or hid them in basements and village storage sheds and then transformed or retransformed themselves into political parties at the end of the war, Hizbullah maintained its weapons, with the blessings of most Lebanese, in order to fight the ongoing Israeli occupation in the South" (Deeb and Harb 2013, 39).[17] Lebanon's long war thus marked merely the group's first iteration, as it has become, in recent years, arguably Lebanon's most powerful political group, a power increasingly measured in parliamentary seats—the group became a formal political party in 1992—but also in popular support both from and also beyond the nation's Shi'i community, in growing institutional and charitable activities in areas such as media, in social welfare services, and in urban development, as well as in military might and acuity, as demonstrated in the summer 2006 war with Israel.

According to the most reliable statistics, around ninety thousand people lost their lives during the war, with tens of thousands kidnapped or disappeared and assumed dead, nearly one hundred thousand badly injured, and close to a million people, or two-thirds of the Lebanese population, displaced.[18] But figures like these are cold and abstract and in fact do little to convey the injuries, devastation, and suffering of war. I had interviewed May, a woman in her fifties, several times before she raised the topic of the war. When talking about how she keeps in touch with her adult daughter in Europe mainly by e-mail, she added, "I don't like talking that way, by e-mail, because I can't sit in front of a computer screen for very long. I don't know why, it must be from the war, but I can't concentrate for long anymore. . . . I get anxious." This is just one example of how the suffering and violence of war endure in people's bodies and souls long after the fighting has ceased. As the central theater of the war, Beirut was physically devastated after fifteen years of conflict, and the city's postwar recovery has been an uneven project of physical as well as national reconstruction.

"THEY HAVE REBUILT THE STONES, BUT THEY HAVE NOT REBUILT THE PEOPLE"

After the Lebanese experienced one of the most deadly phases of the war in the late 1980s, an Arab League initiative, backed by the United States, brokered a ceasefire in Lebanon and organized a meeting of the Lebanese parliamentarians in the city of Ta'if in Saudi Arabia. After a month of deliberations, they agreed upon a Document of National Understanding known as the Ta'if Accord.[19] The accord reproduced the sectarian system but changed the distribution of parliamentary seats, which had been divided on a 6-to-5 ratio of Christians to Muslims since 1943, by establishing parity between the two groups and further subdividing among sects based on their estimated population size.[20] Sectarian quotas were abolished in civil service posts, the judiciary, the army, and the police, with the exception of the general directors of ministries, where parity and rotation were to be applied so that these positions would not be reserved to a fixed sect (Traboulsi 2007, 244). Essentially, the war's end brought the restoration of civil order, the reestablishment of a central government, and the disarming and disbanding of militias, but militia members were in fact absorbed into the army and police force and the sectarian framework of the political structure was preserved.

Unlike other postconflict nations, Lebanon did not undergo any kind of officially sanctioned truth and reconciliation process;[21] as an alternative, the government issued in March 1991 a general amnesty for war crimes that was intended to move the nation forward and away from the war. Moreover, because so many of the leading postwar political figures both within and outside of government were active protagonists and participants in the war,[22] they had little interest in launching an inquiry into the nation's violent past, and none have accepted personal responsibility for any suffering or misdeeds. This official politics of forgetting is also demonstrated by school history textbooks that stop narrating the nation's history in 1946, with the withdrawal of the last French troops.[23] In short, a telling of Lebanon's war, at least in any official capacity, has not yet taken place.

Although some Lebanese emphasize the need for collective forgetfulness by stressing that delving into the past drains the energy and resources necessary for building a new civic order,[24] others, like cultural and literary critic Saree Makdisi, argue that the "general reluctance to engage systematically with the war and embark on a collective historical project to digest and process its memories and images is partly a matter of public policy and partly a matter of a widespread popular will to deny" (2006, 204). However, by arguing that an absence of what he calls "memory culture" at the state level does not tell the full story of how Lebanese have engaged in the work of remembering the long war, Sune Haugbolle (2010) provides a critique of this notion of collective amnesia. His research highlights instead the ways in which Lebanese "memory makers"—artists, intellectuals, architects, and the like—have worked against the state-level processes of amnesia about the war through their cultural production.[25]

At a more everyday level, the phrase "they've rebuilt the stones but they have not rebuilt the people,"[26] which several of my respondents used, registers a popularly held sentiment about the power holders' attention to rebuilding infrastructure and reviving the economy rather than to the postwar project of social repair; that is, Beirutis understand well that the charged social and political landscape of the present reflects the damage done by a recent and volatile past, in which war was waged not only by foreign armies but also by "intimate enemies" (Theidon 2013, xiii): neighbors, co-workers, and friends. As the saying about the stones suggests, a social recovery involving a collective—and state-sponsored—reckoning with the war's past among these intimate enemies has yet to occur.

"WHY LIVE WITH GHOSTS?"

For ten years after the war's end, the physical presence of Syrian army troops, not withdrawn from Beirut and its surrounding areas to the eastern Bekaa Valley region along the Syrian border until 2000 (and later from Lebanon entirely in 2005), was a reminder of the enduring Syrian influence over Lebanon.[27]

On a bright and cool early spring morning in 2006, I walked with Hania, a mother of two in her early forties, uphill along the Corniche seaside boardwalk in Raouché. Off to the side of the walkway was an empty patch of overgrown grassy land where a family with young children was picking yellow wildflowers. In this high-rent part of the city, known for its upscale apartment buildings, luxury hotels, and cliff-side cafes, it was unusual to see an undeveloped no-man's land. In contrast to the idyllic scene of children frolicking in the wildflowers was rusted barbed wire around the area's perimeter and a wooden guard booth. Hania told me that that the area had been a Syrian army position. "It's strange seeing this place empty," she said; "for so long I had gotten used to seeing a Syrian soldier here." We talked about how much had changed in just one year's time, with the full withdrawal of the

Syrian army from Lebanon in 2005 after nearly thirty years, and she commented that "Hariri's assassination changed everything; they [the Syrians] would have been here for another thirty years.... But now they really need to do something with this land right here; they need to change things now that the Syrians are gone." After a pause, she added, "Why live with ghosts?"

Across the physical landscape of this wounded city, ghosts of Lebanon's war do indeed remain. Buildings with bullet and shell holes are a part of the everyday geography, especially in neighborhoods near the Green Line, where fighting was most intense. A badly damaged and hollowed out Holiday Inn that was used as a staging ground for fierce battles during the first years of the war still sits just behind, and in juxtaposition with, the five-star luxury Phoenicia Hotel, which overlooks the Mediterranean. In different parts of the city, plaques, banners, and images commemorate the "martyrdom" of fighters and political leaders who were killed during the war.[28] In this way, making one's way through Beirut constitutes an encounter with physical reminders of the war. These traces of the war configure the city's space both for those who lived through it—just as the seaside area with the empty guardhouse and flowers triggered memories for Hania of the Syrian occupation in the city—and for the youth generation who have no firsthand memories of the war but nonetheless live in a society where memory of the war shapes both the present and ideas about possibilities for the future.[29] And, in downtown Beirut, traces of the war were removed to make way for a multibillion-dollar reconstruction project—which would become one of the largest urban-development projects in the world—that articulated a particular vision of the city.

In the wake of the war's destruction and disorder, Hariri, a billionaire businessman who made his fortune in the Saudi construction industry, seized the opportunity to redevelop and reimagine the city's center. As prime minister from 1992 to 1998 and again from 2000 to 2004, Hariri imagined that Beirut could become a "Hong Kong of the Mediterranean"—a financial and service center that would rival the emerging Gulf cities of Dubai, Doha, and Abu Dhabi. He charted Beirut's postwar recovery along a path of urbanism characterized by the privatization of government services, real estate development, and the enhancement of high-end tourism zones, all in the name of attracting flows of capital.

Situated along the Green Line, which split the city into rival sides, the downtown area was a strategic terrain for militias and, as a stage for the worst of the violence, was significantly damaged during the war.[30] In 1977, during a lull in fighting, the first official plan for what to do with downtown was commissioned by a newly established government agency, the Council for Development and Reconstruction (CDR). The plan was to rebuild the city center along the lines of its prewar layout and to restore its centrality in the life of Beirut in part by ensuring that it retained its prewar character.[31] But the war soon resumed, and later, in 1983, OGER Liban, a private engineering firm owned by Hariri, took over from CDR the project to rebuild downtown and commissioned the development of a master plan. Later that year, and again in 1986, unauthorized demolition destroyed some of the area's most significant surviving buildings and structures as well as several souks (traditional outdoor marketplaces with stalls) and large sections of a residential quarter.[32]

In 1991, the plan for the development of downtown commissioned by Hariri and his firm OGER was made public. It called for the nearly total demolition of existing structures in the city center, required the creation of an island offshore near the port, and made few provisions for property owners and tenants.[33] In 1992, amid parliamentary elections, a series of laws was passed enabling the creation of Solidere, a joint-stock company of which Hariri was the principal shareholder. With the establishment of Solidere, the expropriation of tenants and owners, and the widespread demolition of old buildings, Hariri's reconstruction plan became "irreversible" (Schmid 2002, 245).

Although its inaugural marketing slogan was "Beirut: An Ancient City for the Future,"[34] Solidere in fact revived the French colonial era in the city center. Reconstruction began in what the company refers to on its website as the "historic core," which I described in chapter 1: the star-shaped area of Place de l'Etoile and its radiating streets, which had been developed during the mandate period. The buildings' sandy brown stone reveals tones of orange and pink when the sun hits at certain moments of the day and contrasts with the flat gray that clothes much of the city. Traditional stone arches lead to cobblestone streets, most of which are closed to vehicular traffic. Carvings adorn the buildings and dark wood-slatted shutters and balconies frame the windows on the floors up from the ground level. The area is dominated by retail, much of it offering global and luxury brands priced out of reach for the ordinary Lebanese. Outdoor seating and the pedestrian-dedicated corridors make the area a popular destination for seeing and being seen as well as for families with small children, who pedal around on toddler bikes. In my meeting with Angus Gavin, director of Solidere's Urban Development Division, and other Solidere officials in spring 2005, they noted that this outdoor use of space distinguished Beirut from its "competitors." "The climate in the Arab Gulf," Gavin said, "does not allow for this kind of urban public space."

In fact, it was precisely this type of public space, which feels and looks essentially like a high-end shopping mall, that drew the ire and criticism of so many of Beirut's residents, intellectuals, architects, and artists; they denounced the private takeover of public space as well as the development of an exclusionary city center that reduced individual rights to the city to practices of consumerism and corporate citizenship, as evinced by Solidere's early marketing slogan: "Buy a share in the reconstruction of your city" (Yahya 2007, 247). Many Beirutis I knew shared the sentiments of Lamia, a graduate student, who expressed a sense of loss about downtown's reconstruction: "Downtown doesn't have a Lebanese spirit—look at it. Think about what was there before. It doesn't seem like a Lebanese place at all. It seems like something from France. It's only a place for tourists; it's not for us. It's for the Saudis, and the Hariri-types, the VIPs; you can't afford anything there.... There were souks there before, *real* souks, not like what's there now."

Indeed, as Lamia's comment highlights, much of the popular and scholarly discourse and debate about the Solidere project has focused on its erasure of the past and the ways in which the company's destruction of the downtown built environment was also a destruction of the city's social fabric and memory.[35] Sociologist Nabil Beyhum, for example, has observed, "What the fighting had not managed to destroy of the urban memory and the national heritage, the bulldozers of those reconstructing the city destroyed far more radically" (1992, 52). Critiques of the Solidere project emphasize how the war's urbicidal dimensions—that is, its dismantling of Beirut's heterogeneity into antagonistic enclaves constituted by political, sectarian, and ideological difference—were but a first stage of the city's undoing.[36] As Aseel Sawalha argues, not only did the postwar reconstruction of the city's downtown bring about considerable destruction, but its top-down planning produced socioeconomic ruptures that dislocated the war's displaced one more time, forcing them to create new squatter areas on the outskirts of the city (2010, 132).

The Solidere phenomenon of cultivating a sanitized, uncontroversial branding of the city that avoids references to histories of conflict can be seen in other postconflict cities. In these sites, urban-development projects have not only played a crucial role in plans for economic revitalization and attracting tourists and media attention from around the world but have also been instrumental in efforts to remake the image of cities long associated with violence and war. In Bosnia, for example, the formation of a new postwar multicultural Bosnian identity has been engendered through the reconstruction of the Ottoman-era Mostar Bridge—also situated along a Green Line that divided the city along religious and ethnic lines. In Belfast, which scholars call a once and still divided city, upmarket bars, boutiques, sporting venues, and must-see attractions like the shipyard that built the *Titanic* have been developed to paper over enduring problems of social exclusion and sectarian segregation.[37] In hopes of corralling tourists in "new" parts of the city—exemplified by the "Beirut Is Back" tag line so popular in travel writing following the war—that bear few reminders of past violence, developers and planners in postconflict cities aim to signify urban and national recovery and to market their cities as residential, tourist, and investment destinations.

For Hariri and Solidere, the "geographical axis of competition that pits cities against cities in the global economy" (Smith 2002, 447) meant that postwar downtown Beirut would take on a particularly Dubai-like character. This objective—to achieve a kind of Mediterranean Dubai—was both a reflection of Dubai's rise to prominence as a tourist and business hub during the decades of Lebanon's war and an outcome of the significant increase in tourism to Lebanon by Gulf Arabs, whose travel to the United States and Europe declined after 9/11 because of the inaccessibility of visas and concerns about discrimination and hostility.[38] This vision of Beirut as Dubai in the postwar era left many Beirutis cold however. For instance, during our conversation in spring 2006, architect Rahif Fayad commented that "we're not Dubai, we're not those other cities. We have an urban history, we have urban traditions; . . . this same thinking can't apply, it shouldn't be applied. We have a deep history as a city; we're not a new city, and we're not a Bedouin one." For Fayad, the notion that Beirut could be treated as a tabula rasa and modeled after Gulf cities in the postwar moment was less a vision for the future, as Hariri and his corporation advocated, than a violation of the past.

While many of Solidere's critics decried the erasure of the city's history of conflict, the continuing and deep divides in the Lebanese political landscape in fact came to configure the space in early 2005. Killed in a car bomb

FIGURE 2.1. Tent City: Downtown protestors demand the truth about who killed Hariri and independence from Syrian control, spring 2005. (Photo by author.)

assassination in front of the seaside Phoenicia Hotel on February 14, 2005, Hariri was buried downtown across from a vacant area known as Martyrs' Square, where a statue that commemorates the nationalists hanged by the Ottoman Turks in the early twentieth century is situated. From the time of his funeral and for months afterward, a line of people wanting to visit his burial site extended out into the street, and Martyrs' Square became a tent city for demonstrators with signs demanding *al-haqeeqa* (the truth) about the identity of Hariri's assassin.[39]

As the tenor of politics grew increasingly tense in the wake of Hariri's assassination, tens of thousands of pro-Syrian, pro-Hizbullah protestors gathered in another part of downtown named for the statue of Riad al-Solh, Lebanon's first prime minister after independence. In a speech there, Hizbullah leader Hassan Nasrallah thanked Syria for its support of the resistance against Israeli occupation. Six days later, on March 14, 2005, an estimated million people assembled at Martyrs' Square and called for the Syrian army's withdrawal from Lebanon. On these days, the nation's two most powerful political unions of the post- Hariri era—the

March 8th coalition and the March 14th alliance—were born. As the spring continued, the media-termed and heavily youth-based Cedar Revolution, which supported the politics of the March 14th alliance, seized the area around Martyrs' Square as its own: banners, graffiti, and flags adorned the area around the tents. Journalists from around the world descended, and filmmakers took footage for documentaries. Downtown, in this moment, did not remind anyone of Dubai. Rather, it was a space imbued with Lebanon's enduring political polarization, a site on which Hariri had once cast his vision for the postwar city but on which he was now laid to rest.

NOT YET PEACE

> During the war, our eyes were always fixed on what we were sure would be the halcyon days of the future after it ended. Let the war end, we thought, and all would be well. We would emerge from the abyss into the light. Historical quarrels and divisions would mutate into a harmonious and productive unity based on justice. In this vision of the future, I think, we felt somehow that the best of the past would be preserved, the worst purged by our travails. We had paid a heavy price for the evils of the past, and we deserved a better world. But the future is now, and it is a hard reality, shorn of these illusions. There was to be no reward, after all, for the suffering. (Makdisi 1999, 257)

Written by Jean Makdisi in 1999 in the afterword to her memoir about Lebanon's war, these words also capture the sentiments of most people I knew during my research in Beirut. Although the war may have ended, not only did its effects endure, but other wars also began. Beginning in 2005, amid the emergence of the two competing political camps—the March 14th and the March 8th coalitions—near monthly car bombs exploded in Beirut that killed or attempted to kill political figures and journalists. During summer 2006, in response to Hizbullah's capture of two Israeli soldiers along the border, Israel engaged in a full-scale military assault that affected the whole country but mainly killed civilians and destroyed infrastructure in Shi'i-majority areas: the southern parts of the country, the eastern Bekaa Valley, and the southern suburbs of Beirut. In addition to the more than nine hundred killed, thousands were wounded, and nearly one million were displaced from their homes—one quarter of the country's population.[40]

Entire villages in South Lebanon were flattened during the attack as were whole neighborhoods in south Beirut. Unexploded ordnances from the war continue to threaten the population.[41]

The rancor between the country's two main political groups increased during 2005, worsened after the summer 2006 war, and developed into an economic and political deadlock in 2007. Contention surrounding who would succeed Emile Lahoud as president deepened the rift, and in May 2008 civil war threatened when street battles and the seizure of several West Beirut neighborhoods by Hizbullah fighters ensued after a standoff between Hizbullah and leaders of the March 14th coalition. The fighting ended only after a deal was reached between Hizbullah and government officials. Since the intensification of the Syrian conflict in late 2011, escalating political tensions, a humanitarian crisis created by the arrival of nearly one million Syrian refugees, and armed confrontations in various parts of the country, Lebanon is once again being described as "pushed to the brink" (Mudallali 2013).

Given this context, applying the periodization of "postwar" to contemporary Lebanon must be called into question. Like Belfast and Bosnia, political, spatial, and social divisions induced by past civil conflict are relevant aspects of urban life in the present. But, unlike these other cities, Beirut is shaped not just by a past war but also by ongoing armed conflicts fueled and sustained by transnational networks that extend beyond the nation and region. I thus share anthropologist Isaias Rojas-Pérez's concern about the usefulness of postconflict as a category for the analysis of societies that have experienced conflict. "Before" and "after" scenarios, he observes, can obscure the specific ways in which violence repeats itself (2008, 255). In this regard, it is important to consider how conditions of war both past and present shape everyday life in the contemporary space and society Beirut.[42] The shelled Holiday Inn that housed militias during Lebanon's war, for instance, sits just down the street from where a car bomb that exploded in front of a downtown office building killed several people, including former Finance Minister Mohammad Shatah, and injured scores of others in December 2013. Sites of political violence from various periods of history—from decades ago to very recently—commingle in the city's urban environment.

Political violence has been one of the key power geometries, along with modes of privatization, shaping the city's space in the modern era: spatial organization, access, and mobility have not only reflected conflict among political groups but have also been tools used by political groups to gain and expand power.

In the next chapter, I explore how residents move through and understand this precarious urban space, often with hopes for a different and peaceful future not yet realized.

3 · POLITICS AND PUBLIC SPACE

"The Green Line, you know, it still exists in people's minds." When Mounir, a recent university graduate, spoke about the Green Line, which divided Beirut into a predominantly Muslim West and mainly Christian East during the long war (1975–1990), he drew attention to the fact that the spatial repercussions of the war, now more than twenty years in the past, endure. The conflict played out in public space and created a polarized city in which neighborhoods that were identified with particular sectarian groups carried assumed affiliation with certain political parties and ideologies. In this chapter, I explore how, in the 2004–2006 period of renewed political conflict and violence, ideas about and practices of mobility were shaped by—and also shaped—this fractious political geography. Here, I aim to present a picture of what residents' uses of the city can tell us about the workings of political sectarianism in the city.[1]

Like other divided cities, Beirut's spatial polarization has been connected with issues concerning the very legitimacy of political structures and the struggle over access to governance and state institutions.[2] In one sense, this characterization aptly describes the events of the 2004–2006 period, when struggles between the two main political camps, the U.S.-backed anti-Syrian March 14th alliance and the Iranian-backed pro-Syrian March 8th coalition, involved competition for power and influence both within and outside the government by, for instance, gaining seats in parliament and expanding social-welfare institutions.[3]

At the same time, this understanding of the city's space as being polarized in connection with issues of control over and access to the formal political system does not adequately capture the ideological tenor of everyday geographies in Beirut. For example, in the days following the March 8, 2005, assemblage at Riad al-Solh (a square in downtown Beirut) of an estimated half a million people who gathered to support Hizbullah and the pro-Syrian stance, women at the small fitness center I patronized urged me to join them for the counter-protest on March 14th. The counter-protest was to take place in another downtown site, Martyrs' Square, where anti-Syrian protestors demanding *al-haqeeqa* (the truth) about who killed Rafiq Hariri had set up encampments. Assuming that I was politically aligned with the U.S.-supported March 14th group, Karine, one of the owners of the gym, was resolute in her appeal: "You have to come," she said. "We all have to go down there and show them [Hizbullah and its supporters] that they are not Lebanon; *we* are Lebanon." In the weeks, months, and years following the two March 2005 protests, this us-versus-them sentiment of rival political ideologies came to prevail over the space of downtown so much that by 2007, as Ward Vloeberghs describes, "one literally had to 'choose sides' in order to access the restaurants and retail in the city center; either entering through an area of tents installed by March 8th supporters or by passing by Hariri's burial site adjacent to Martyrs' Square and thus paying respects to the March 14th camp" (2012, 160). In this way, while political polarization is not a phenomenon unique to the space of Beirut, the ways in which politico-ideological issues spill over into space and are set off by domestic and regional geopolitical events gives everyday use of the city particular meaning.

I begin the chapter by outlining some of the spatial dimensions of political sectarianism in the city and then focus on one specific historical moment, that of the period of protest in early 2005 throughout the Arab world against the Danish newspaper cartoon depicting the Muslim prophet Mohammad, to illustrate how events can trigger violent responses that abruptly transform urban spaces and create dividing lines between them. Next, I draw on my research with young people whom I define as socioeconomically "privileged"[4] because of their educational backgrounds and levels of disposable income to show that although leisure pursuits in particular can encourage mobility to different parts of the city, sociospatial boundaries that constrain this mobility can also be reinforced during moments of heightened political tension. Conceiving of these young people as social agents and competent urban actors in their own right, I consider what their movement in the city reveals about the spatial aspects of social class and political sectarianism. To begin, however, it is important to present a picture of the appearance of politics and sect in the city's public space.

POLITICAL SECTARIANISM AND BOUNDARIES IN URBAN SPACE

Although few neighborhoods in Beirut are entirely homogenous in the sectarian affiliation of their residents and very few areas are completely mixed, expressions of these affiliations are a part of the visual and acoustic fabric of the city.[5] For instance, while walking through neighborhoods near my apartment when the Islamic calendar reached its final month (*dhou al-hija*), I would see colored streamers draped across verandas in celebration of those who have returned from the pilgrimage (*haj*) to Mecca, one of the pillars of Islam; I came across glass-encased boxes along streets in parts of the majority-Christian areas of eastern Beirut that held Virgin Mary figurines; throughout the city, the sounds of ringing bells and calls to prayer emanated from churches and mosques. Once, while in the mainly Shi'i Dahiya, I walked past a Husseiniya, a gathering place and site of the commemoration of Imam Hussein's martyrdom that is erected during Ashura, the Islamic month during which Shi'a mourn the martyrdom of Hussein in the seventh century. Expressions of sectarian and religious affiliation also take sartorial forms that are publicly visible, from necklaces with dangling crosses to the diverse range of Islamic head coverings and outer garments worn by both men and women. These kinds of sights, sounds, and symbols do not necessarily function as borders keeping people out or as gates keeping people in, but they can and do territorially express sectarian belonging.

As I described in the first two chapters, politics and sect come together in Lebanon as religious sect has historically served as the basis for political identity and representation. Sectarian affiliation is no longer indicated on national identity cards, and, since 2009, it is not required on civil registry records, but the sectarian dimensions of the power-sharing political and electoral structure have remained unchanged since the signing of the

Syrian-Saudi-sponsored Ta'if Accord, which ended the protracted war in 1990.[6] State institutions continue to be under the helm of individuals with links to elite families from the different sectarian communities, and the state's allocation of welfare resources, rather than being distributed on the basis of need, is tied to both the sectarian distribution in the country and to the rewarding of political activism that emerges from sectarian-based political groups.[7] In short, sectarian identity remains salient in matters of politics, civic life, and even livelihood, as connections to and building networks (in Arabic, *wasta*) with one's sectarian community and its associated elite families continue to enhance individuals' access to social services and provide avenues for socioeconomic mobility.

Expressions of sectarian identity in Beirut's built environment can also be politicized. In 2004, during a conversation about the ongoing downtown reconstruction with Howayda al-Harithy, a professor of architecture and urban planning at the American University of Beirut,[8] she drew my attention to what, in her understanding, was a sectarian dimension of the area's built environment. Speaking about the building of the Mohammad al-Amin mosque in the downtown area, a blue-domed behemoth of a building, she thought about the mosque as a public demonstration of Sunni identity, the sectarian affiliation of its developer and financer, former prime minister Hariri. "Building it in such an Ottoman style, there are no other mosques that look that way in Beirut, . . . and the Ottomans represent the height of Sunni power in the Middle East. That is the connection that comes to my mind."[9]

As founder and part owner of Solidere, the company at the head of the downtown area's postwar redevelopment, Hariri, perhaps unsurprisingly, intended the mosque not only to proclaim his role as the *za'im* (communal leader) of the Sunni community but also as a tribute to Sunni heritage in the region. For Sawsan, a university student, the mosque was overbearing not only because, as she put it, "it takes up so much of downtown" but also because it was understood as a political claim to space cast in religious terms: "I don't think this," she said, "but I do know people who are annoyed by it [the al-Amin mosque]; they drive by it and say, 'They just put that here to show that it's a Muslim country.'" In this sense, while Solidere was active in efforts to recover the downtown area's many religious landmarks, both Muslim and Christian, in the wake of the long war—in 2006, for instance, seventeen religious structures were being restored or reconstructed in the area[10]—the expanse of the al-Amin mosque along with Hariri's identification with it and his steering of downtown's development were understood by some as a political sectarian assertion. Just as Lucia Volk (2010) describes in her examination of Lebanese public memorials and monuments, the built and spatial environments are important venues through which elites in Lebanon lay claim to the space not only of Beirut but also of the nation as a whole.

Iconography also mobilizes political sectarianism in the spatial realm as portraits and photographs of political leaders visually inscribe relations of kinlike loyalty between members of a sectarian community and their "custodian."[11] As a common phenomenon in the symbolic practices of Middle Eastern politics,[12] these images, fastened to building walls, affixed to posts, and hanging across streets, "amplify the leader into a mythical hero by idealizing him as the protector of his community and its sectarian interests" (Maasri 2009, 57). Alongside the cult of personality created by these images, other types of visual displays—flags, murals, graffiti, and banners—mark out political turf and territorialize neighborhoods as belonging to particular political parties.

In Beirut, sometimes it is not even the symbols or slogans on the flags and banners that express political affiliation, but their colors, for the most powerful political parties carry associations with one or more colors. During spring 2005, for example, amid the protests following Hariri's assassination, it was common to see people with light blue ribbons pinned to their shirts in the style of the red AIDS awareness ribbon. The light blue color signified support for Hariri's Future Movement political party, the leading member of the March 14th coalition. Throughout the city's public spaces and along highways and roads in other regions of the country, these colors signify the various political teams and the leaders to whom their supporters profess allegiance. In the U.S. tradition of baseball trading cards, the marketing potential of these associations was seized on by the Safina Group, which began selling, in June 2007, "Parties and Colors" collectible sticker albums and stickers depicting the leaders and flags of the Lebanese political parties.

In the predominantly Shi'i southern suburbs of Beirut, the yellow and green flags of Hizbullah adorn the streets, as an Islamist militant politics of resistance to Israeli claims and Western intervention in the Middle East prevails over the landscape. Dahiya, as the area is known (literally, "the

Politics and Public Space 61

FIGURE 3.1. Lebanese "Parties and Colors" albums and stickers. (From www.safinagroup.com, captured 1/11/08 by author.)

suburb" in Arabic), is an assemblage of multiple municipalities and neighborhoods.[13] In the early twentieth century, the land where it is situated was rural, but by 1970 (Deeb 2006, 47) it had become part of Greater Beirut, mainly because of the wave of rural-to-urban migration in the 1950s and 1960s, which I describe in chapter 1. During the years of Lebanon's war, Shi'i refugees from the northeastern suburbs, the south, and the Bekaa Valley arrived in the area, and these consecutive waves of migration altered the sectarian makeup of the southern suburbs from being a mix of Shi'i Muslims and Maronite Christians to being predominantly Shi'i. Although, in many ways, the neighborhoods in Dahiya are unexceptional, resembling other neighborhoods in Beirut, especially working-class ones, the area is set apart by its population density and the morphology of the built environment.[14] Although, as Mona Harb writes, Dahiya is just another Beirut neighborhood managed "by a sect-based political actor," by reserving for itself the right of armed resistance to Israel—a claim some Lebanese see as a threat to the legitimacy of the state—the group constitutes the southern suburbs as a realm distinct not only from other parts of Beirut but from the rest of Lebanon (Harb 2007, 13).

In her ethnography about expressions and understandings of piety among the Shi'a living in Dahiya, Lara Deeb describes the area's public space as a "mix of politics and piety with temporal, visual, and aural textures that contribute to a collective sense of community" (2006, 48). Dahiya is marked as a pious space by commemorative images associated with the Islamic resistance, such as martyrdom posters of fallen soldiers (which also appear in other neighborhoods) and other spatial forms of sacralization such as banners and flags that mark the cycle of the Islamic calendar, residents' clothing and conduct, as well as leisure and entertainment venues that adhere to Islamic notions of respectability.[15] Hizbullah also effectively maintains its own security forces in Dahiya, and foreign researchers I knew had been stopped and questioned by members of these forces about their reasons for coming into the area during the tense 2004–2006 period. However, in other neighborhoods, too, there are visible forms of extrastate security whereby local residents (mostly male) appoint themselves protectors of their politically sectarian neighborhood, reconstructed generally as a sectarian territory, and question those perceived to be outsiders about their reason for being in the neighborhood.[16]

Friends and acquaintances of mine who were politically aligned with the March 14th coalition, which opposed Hizbullah, were often surprised and sometimes concerned when they learned that, at one point during my fieldwork, I was regularly going to meet with a research participant who lived in Dahiya. In the contentious 2004–2006 political atmosphere, which drew sharp lines of enmity between supporters of the March 8th (Hizbullah and allies) and the March 14th groups, Dahiya, because of its identification with Hizbullah, represented a definitively adversarial territory for the March 14th coalition. In this way, I found that among people I knew avoidance of Dahiya and even negative associations with the area had mainly to do with their own political positioning. But Lebanese are reluctant to go to Dahiya for other reasons. Service taxi drivers would often refuse me as a passenger when I was headed there or ask for me to pay a double or triple fare because of its traffic congestion. Other Lebanese have trepidations about going there because of an apprehension that stems from unfamiliarity about what it would mean to be in a Hizbullah-controlled area.

Once, when finishing up an interview at a Lebanese NGO, I asked two members of the staff how to take public transportation to the airport area as I had to pick up a package being held at a post office near there. To go such a distance from the offices of the organization in Hazmieh, a suburb near the Damascus Highway, I knew that I would probably have to take more than one bus or service taxi and wanted to figure out where I should make transfers. The staff discouraged me from using anything other than an expensive private taxi that would take me directly to my destination because, they warned, I would have to go through and perhaps stop in Dahiya en route. "I'm sorry," a staff member named Aline said, "I just can't let you take a service there. . . . I wouldn't let my own sister do that! We'll call a private taxi for you." When I asked her why she thought I should not go through Dahiya, she said, putting an end to the conversation, that the area was "unsafe." Aline's efforts to discourage me from going through the area illustrate the kinds of boundaries that mark Beirut's urban space.

These boundaries, I learned, were spatial in more than one way, for they had to do not only with the delineation of different urban territories but also with the mode of transportation itself. At one point during my fieldwork, after having ridden on both of them many times, I asked a Lebanese friend why there were two different types of larger public buses that charged the same fare: the red and white Lebanese Commuting Company (LCC) buses, which had a more official feel replete with a machine that automatically dispensed a receipt after you handed the driver your payment, and the less formal white buses, whose drivers were regularly changed along the route. "Well," she replied, "I've heard that the LCC company is owned by Hariri and the white buses are run by Berri." Nabih Berri, the speaker of the parliament, Lebanon's third most powerful political position, is the leader of the political party Amal, which is associated with the Shi'i community and is a member of the March 8th alliance, which opposes the March 14th coalition, led by the Hariri's Future Movement party. In other words, Berri and Hariri are members of opposing political camps. When I asked others the same question, a few respondents said they were not sure who owned the white buses—many thought that individual owners and not Berri operated them—but they too presumed that the LCC line was one of the Hariri family's bottomless assets.

The veracity of the notion notwithstanding, what is striking about the comment about one bus company being owned by Berri and the other by the Hariri clan is the way in which it depicts the public bus system itself as a field of competition between two different political factions with attachments to particular sectarian communities. This notion of the "sectarianness" of the very means of mobility emerged too in people's talk about service taxis. For instance, when speaking about the routes service taxi drivers choose to circulate along in their hunt for passengers, Amal, a secretary in her fifties, commented that "it is true that if they [service drivers] are Muslim, then they drive through the western part of the city, and if they are Christian they stay mostly on the eastern side—this is from the war [*min asl al-harb*]. They are just used to driving around their part of the city; ... it's a habit from the war." Amal's idea that service drivers prefer to drive through areas with which they feel a sense of political sectarian belonging emerged in many of my other conversations with the city's residents. When I asked several service drivers while riding as a passenger in different parts of the city about the routes they took, most described adhering more or a less to a loop through areas situated on either the eastern or western sides of the city. One driver I spoke with, a man who said he was sixty-seven and had been driving a service for forty-five years, talked about the route he stayed on, which started from Bourj al-Barajneh in Dahiya and ended in Hamra and then back again: "Of course I will take someone where they need to go in Ashrafieh [in the east] or wherever," he replied in a frustrated tone to my question about driving off the route he described. "But then I go back to my route." In this way, ideas about and practices of public transportation in the city are shaped by past and present everyday geographies enmeshed in notions of politics and sect. The politics of sectarianism are made public and visible in the city by various means, from the hanging of flags and banners in support of sectarian-identified political groups through the interpretation of monumental architecture as a political sectarian claim on parts of the city to understandings of public transportation as being in the pockets of leaders of sectarian-affiliated political parties. In these ways, Beirut is territorialized by—and the city's space is made an active agent in—the processes of political sectarianism.

THE ANTICIPATION OF VIOLENCE

What is at stake in everyday navigations of this political sectarian landscape I have just described? For one, talk about politics is an ordinary topic in day-to-day life, even among strangers getting around town. Service taxis, the most widely used form of public transit, nurture this discourse as they serve as a kind of moving town hall in which Lebanese from diverse backgrounds come into contact, and conversation, with one another. Karim, a student at the American University of Beirut, said he usually avoids taking the front seat of a service taxi if one in the back is open: "It's just annoying because . . . if I take the front seat, then the driver ends up talking to me the whole time, and always about politics. I don't know why, but there's this thing about *shabab* [young men/young people]: the drivers think that we're the ones they can talk to about anything!"

Although not all Beirutis are politically engaged, politics, as Karim put it, cannot be avoided, even when just making your way around the city. When Nadine, a recent college graduate, talked with me about her participation in the spring 2005 protests against Syrian involvement in Lebanon, I asked her if she could remember when she first began taking an interest in politics. She looked at me curiously. "Politics is simply a part of being Lebanese," she said succinctly. Indeed, just as most Lebanese newspapers and television stations are aligned with a particular political perspective—if not directly produced and financed by a political party—the news radio station a service taxi driver chooses to listen to communicates to passengers something about his political orientation even before any conversation in the vehicle begins. In the post-2004 era especially, when fault lines between different political and ideological camps became increasingly entrenched, a taxi driver's choice to listen to a news broadcast on the Hizbullah-sponsored al-Nour station, the Hariri-led Radio Orient, or the Phalange Party's Voice of Lebanon,[17] for instance, could set the tone for any potential political conversation in which coming and going passengers might participate. While politics might play a dominant role in public discourse between strangers in many other parts of the world, in Beirut the stakes of politics are immediate and ever-looming not only because of the existence of what Sune Haugbolle (2010) refers to as "multiple memory cultures" surrounding Lebanon's long war of the past but also as an outcome of how politics regularly change the course of one's daily activities, and feelings of safety, in the present.

During the 2004–2006 period, when political tensions ran high amid the series of assassinations and the consolidation of the opposing March 8th and March 14th coalitions, even events that occurred outside Lebanon could serve as flashpoints for protest and violence that would, if only for a time, transform the city's space. After first being published in the Danish newspaper *Jyllands-Posten* in September 2005, cartoons ridiculing the Islamic prophet Muhammad were reprinted in January 2006 in newspapers in more than fifty other countries. Protests against the cartoons, some of which escalated into violence—with protestors destroying European buildings and burning flags, and police firing into crowds—ensued throughout the Arab and Muslim world. On February 4, demonstrators in Damascus stormed and set afire the Norwegian and Danish embassies. A day later, on Sunday, February 5, I gathered with friends for brunch at an apartment in Zarif, a neighborhood in central Beirut, on the western side of the city, just south of one of Beirut's only public parks, Sanayeh Garden. A friend received a text message saying that we should turn on the television to see what was happening just a mile and a half east of us. We turned it on. Live news coverage showed men running and shouting through Ashrafieh and surrounding neighborhoods on the eastern side of the city that are generally identified—as they had been during the civil and regional war—as "Christian." Flames shot from garbage dumpsters and vehicles, debris lined the streets. After attacking the Austrian and Danish consulates, the protestors went on to destroy property along the adjacent streets.

From the television, it looked awful. As if Beirut would burn. The reporting on the station we watched identified those participating in what was happening as Palestinian, Syrian, and Shi'i. The men were thus understood to be Muslim, and their acts were thus defined as anti-Christian aggression. As we watched the live news coverage, we heard cars driving by with loud music blaring and voices shouting through megaphones their support for the protestors. It was uncanny to be watching the events unfold at the same time we were hearing these caravans driving through the neighborhood because there was no time lapse between the protests taking place and the deployment in the streets of those who supported the protests; the two events were coterminous. This kind of temporality, with actions and reactions overlapping and no time in between, was one of the disquieting aspects of the city's political climate as violent events could spin off of one another very quickly, gaining fervor, leaving residents to suddenly and quickly take up positions of defense and seek out places of safety.

We worried that the violence would continue throughout the night or provoke the young men who lived in the neighborhoods where the protests and violence were taking place in Ashrafieh to seek reprisals by coming over to the western side of the city, where most of us lived, an area still often conceived—as it had been during the long war—as Muslim: "We either need to stay here all afternoon or get home now," my friend Siham warned, her eyes fixed on the television, "because next these guys [from Ashrafieh] are going to head to Tariq al-Jedeideh [a majority Muslim neighborhood] or even come over here to do the same or worse."

We were not the only ones concerned about such a possibility. Walking home that afternoon along mostly empty streets, I saw thirty or so army soldiers stationed in a vacant lot behind a mosque. Their guns were collected in a mass in the middle of the lot propped upright while the soldiers sat on the ground or stood around the perimeter of the lot. It looked as if they were waiting for something. They might have been stationed there to protect the mosque from possible retributive damage or perhaps to defend this part of the city from attackers.[18] The mosque was situated along Spears Street, a major thoroughfare running from western to eastern Beirut. In other words, it was the most direct route for anyone from the eastern side of the city seeking retribution in the neighborhoods of western Beirut. They could come across the line that had divided the city into Muslim and Christian sides during Lebanon's long war and that, now and again, still seemed to split Beirut in two.

On that afternoon, the near empty streets and the soldiers' presence signaled an anticipation of violence among city residents as well as the state. This anticipation, as Pradeep Jegnathan observes (2002), is embedded in past experiences of public and political violence. Soldiers are deployed quickly, and residents hurry home to safety with efficiency because they have had cause to do so before. For me, the emptiness that day was reminiscent of the hours that followed news of former prime minister Hariri's assassination in February 2005, when all of Beirut seemed to shutter in a mere hour after news came of the blast that killed him and his caravan. But the emptiness too recalled Jean Makdisi's writing about life in the city during the civil and regional war: "Suddenly, gunfire exploded on the street outside, and dozens of people scurried into the café to take refuge from the battle. . . . The street was suddenly deserted. Beirutis have broken all records for getting out of the way on time. It is incredible to see how quickly a street swarming with people can be transformed into ghostly emptiness. Shopkeepers close their doors and pull down their iron shutters, mothers scoop up their children and run, vendors scuttle away with their carts, and after an even more than usually furious beeping of horns, the traffic jam evaporates in no time at all" (1999, 86).

In this way, temporal fields of both remembrance of the past and anticipation of the future overlap in Beirut's fraught political landscape as the city is a terrain where residents' experiences of earlier political violence become entangled with concern about the violence that is to come. Violence is anticipated as an outcome of deductions made by residents as well as by state and privately hired security personnel and can be based on interpretations of the current level of political tension that is conveyed through the daily tenor of news broadcasts as well as through the intensity and vocabulary of political debates by news commentators.[19] These calculations of danger are also made from what city dwellers detect to be an intensification of security, which is demonstrated through, for instance, the rapid deployment of armed soldiers to a particular locale, as was the case on that day of unrest over the Danish cartoons in February 2006. Through this beefing up of security in public space, residents figure out that the members of the security apparatus are gearing up for a potentially volatile situation, and, as a result, residents anticipate not only the possibility of violence but also the necessity for altering their daily movements through the city.

CROSSING BORDERS

In 2005, amid the growing rancor between the March 14th and the March 8th political camps and the withdrawal of Syrian troops from Lebanon at the end of April 2005, a series of bombs exploded around Beirut. The bombing targets were mainly journalists and political figures allied with the March 14th coalition and its anti-Syrian stance, but some were also detonated in industrial and commercial areas of the city with predominantly Christian populations. In this sense, the pattern of violence that emerged in spring 2005 was one that overlapped political and sectarian identification as both people and places identified with the March 14th coalition of leading Sunni Muslim and Christian political parties were targeted.

This mapping of the space of the city concerned Layla, a mother in her early forties, and influenced her decision making about where and when her teenaged children, a boy and a girl, would be allowed to socialize: "I don't want them going to ABC," a newly opened high-end shopping mall in Ashrafieh, a predominantly Christian neighborhood where the Danish cartoon protesters had unleashed their fury. "No one knows for sure where the next bomb will go off, but ABC is a likely target," she thought. "I would rather they [her children] stay close to home; they can do shopping here, they can just go to see a movie here, in this part of the city. No bombs have gone off here." At the ABC shopping center, women's handbags were checked and men were patted down by security officers stationed at the mall's entrances, while at the Dunes shopping plaza, around the corner from Layla's residence in Verdun, an affluent neighborhood in western Beirut, these kinds of security measures were not a regular practice. But the Dunes shopping complex, built in the 1990s, was a far less desirable recreational destination for Layla's son and daughter, who, like many other young people, wanted to go with their friends to the newest hangouts with the most up-to-date cinemas and retail outlets regardless of where in Beirut they were to be found.[20]

While Layla set parameters for her children's movements through the city according to her construal of the potential dangers created by political violence, some young people had a different attitude about Beirut's geography of risk. In 2006, I arranged to meet Rana, a student at a French-language Jesuit university. At a McDonald's near campus, Rana spoke about her notions of safety in getting around the city. When I asked her about whether she had changed her routes or destinations since the spate of bombings had begun a year earlier, she gave a remarkable response: "No, I haven't changed anything; I still go out. But, actually, it's funny, after something has happened, I mean after there has been an explosion somewhere, that's when I feel like I can go to that place since it's already had an explosion. My friends and I feel like, 'Well, they've already put a bomb there, now it's safe to go out there!'"[21]

Rana, a young woman from an upper-middle-class background who lived in the northern suburbs of Beirut, was furnished by her parents with a car, spending money, and consent to go out pretty much when and where she pleased—usually, she said, this was to meet up with friends at cafes, restaurants, and pubs. For her, the city seemed not to be mapped so much by the affiliations and outcomes of political sectarianism as by certain kinds of lifestyle practices; she and her friends sought out trendy places that were distinguished by their American and European branding.

A common, if not clichéd, summing up of Beirut highlights its paradoxes: a coastal urban playground abounding with nightlife, boutiques, and joie de vivre coexists with the threat of violent political conflict.[22] For Rana and her friends, living with this paradox meant developing a rationale for judging places or areas to be safe. Rather than being shuttered inside by the wave of bombings, they sought out locales that had already been the site of an attack, on the presumption that sites were rarely targeted twice. This rationale, as a means of asserting control over the violent phenomena produced by political instability and conflict, is an example of the way in which some Beirutis make sense of and respond to the everyday forms of public violence that surround them. And for residents with the means, like Rana and her friends, middle-class consumer lifestyle practices perhaps even serve as a kind of salve for the anxieties engendered by living with the anticipation of violence, as they seemed to do for Mounir. "The thing is," he related, "we're not going to let politics get in the way of going out. That's what they want, they want to stop everything and shut life down, . . . but we want to live, we want to have fun. Even with the bombs, we're not going to stay home."

Moreover, for middle-class youth with the disposable income to support regular leisure activities, the city's divisive political geography was no match for having a good time. Young people who had the money to spend on eating out at places like the American-style Roadster Diner, seeing the latest Hollywood films, shopping at European-owned retail clothing chains like Zara, Mango, Jack Jones, and Vera Moda, or going to one of Beirut's numerous nightclubs with US$20 entry fees and US$12 cocktails would travel to any and all of the city's neighborhoods to seek out these experiences with little thought given to the political or sectarian association of the area to which they were headed. Mounir, who left Lebanon for Denmark as a child but had recently returned, identified himself as a Muslim but said he would often go out with friends to Gemmayzeh or a neighboring area, Monot (nightclub areas in predominantly Christian neighborhoods). "I don't avoid those areas because they are Christian areas because this is where everyone is going out right now; . . . so this where my friends and I want to be. It doesn't matter that it's a Christian area." In this sense, the everyday mobilities of the middle-class youth I spoke with seemed to be motivated mainly by cultural and consumer sensibilities and leisure practices, and these sensibilities and practices led them to cross the territorial boundaries of the city's political sectarian geography. In the words of Mounir, these were young people who wanted to be "where everyone was going out" regardless of a destination's political and sectarian identification.

I also met young people who sought out parts of the city that represented a melding of boundaries as a matter of principle. Zeina, for example, an American University of Beirut student from a family that left to live in Australia during the long war and returned to Lebanon in 1995, told me that she preferred to go out in Hamra (a neighborhood and consumerist paradise in West Beirut) because "it feels the least sectarian." "What do you mean by sectarian?" I asked. "I mean," she said, "Hamra just feels like a neighborhood that doesn't belong to any one sect, it's so diverse; . . . that's why I prefer it. I used to like going downtown too, when all the reconstruction was going on and there were ruins. . . . I used to go there, but then that became a sectarian place. I don't mean it became known as a place that belongs to a certain sect; I mean it became where people go to show that they are a part of this sect or that sect. So now I don't like going there. I don't know, when I go out, I'm looking to get beyond the problems we have in this country. I don't want to be reminded about them."

As a means of trying to move beyond what she saw as the sectarian-ness of the city's public spaces, Zeina sought out areas where she thought sectarian sentiments and identifications were less demonstrable. For Zeina, choices about leisure had to do with more than locating the trendiest and most popular spots; they involved locating spaces that aligned with what she described as her "antisectarian" orientation. Her concerns about how to navigate Beirut's politicized space are not unlike those of the city's pious youth that Deeb and Harb (2013) write about, youth whose day-to-day negotiations about where to go out in public and socialize involve efforts to have fun while striving to adhere to religiously informed moral values. In other words, the pursuit of fun, for some of Beirut's young people, involves satisfying both consumerist and recreational desires as well as those relating to moral, religious, and political views.

The urban mobilities of the privileged young people I spoke with, all of whom had been or were full-time university students with the financial backing of their parents, showed traversals of Beirut's political sectarian geography. With their shared cultural sensibilities situated around a globalized "cosmopolitan cool," it often seemed that the class, rather than the political or sectarian, dimensions of their mobility practice were primary. These crossings resemble those of the upper-class youth in Cairo whom both Mark Allen Peterson (2011) and Anouk de Koning (2009) describe. In both Beirut and Cairo, class-specific leisure destinations like upscale U.S. and European coffee shops, such as Starbucks and Costa, and local outlets styled after these global chains are sites through which privileged young people "express and construct their cosmopolitan identities" (Peterson 2011, 169). For most young Beirutis I met who had money to spend, decisions about where to go in the city were thus motivated by lifestyle choices shaped by class and cultural positioning.

In some ways, this was an unsurprising finding about a group of young people whose geographic mobility was a constitutive feature of their class position. Lesser privileged youth—those, for example, whom I met as a volunteer with the Lebanese agency running the U.S. State Department's English-training Access program for teens from economically disadvantaged areas of the country—had traveled minimally within Lebanon and none had traveled outside the country. In fact for some of the kids I spoke with, participation in the Access program brought about their first-ever visit to Beirut, even though they were from villages just four hours away from the capital. Rana and her friends, however, moved through not just Beirut but also the country in pursuit of various leisure activities from "clubbing" to visiting beach resorts in the summer and ski resorts in the winter. They also had opportunities to leave Lebanon, as young people whose family's financial capital afforded them the possibility of being approved for a visa to go *barra* (outside Lebanon). They went for a touristic vacation or a visit with a relative to countries in Europe, North America, and the Arab Gulf. Sabine, for example, a recent graduate of the city's Jesuit French-language university, talked with me about her and her friends' study-abroad experiences in Grenoble, France, while Noura, another university student, spoke about having visited her relatives in Canada on several occasions.

The mobility of the privileged youth I spoke with was also unconstrained in several other ways. First, they bore little, if any, financial responsibility for themselves or their families. Unlike laboring youth, whose movements around the city were related primarily to livelihood—mobility that has to do with seeking employment or heading to and from work, for instance—the mobility practices of the young people I discuss here were motivated by leisure; none were expected to be financially independent or contributing to family household expenses. While many were or had recently been full-time university students, they did not talk as if there was an expectation on the part of their parents that this financial dependence would shift soon after graduation. Second, for these youth, both the young men and the young women, the rite of going out with friends, even until the early morning hours, was one socially sanctioned and financially supported by their families. Finally, their mobility and public behavior were unconstrained by state authorities, in contrast, for example, with other young people, like Syrian migrant laborers, who endure extra scrutiny by the state police and armed forces as they move around the city.[23] In short, for the young Beirutis I spoke with, various forms of geographic and urban mobility were a critical feature of their class habitus: being spatially mobile was in fact a way of being middle and upper class in values, disposition, and lifestyle.

THICKENING BORDERS

In the 2004–2006 period, there were moments, one that even stretched for months, in the period following former prime minister Hariri's assassination, during which the tides of everyday life were altered by a sense of doom and uncertainty wrought by the series of bombings and the unstable political situation. It was during this time that Nicole described how "right now, because of what's going on, when I get into a service taxi, I just feel more comfortable if I see that the driver has a cross [taxi drivers often hang religious icons from their rearview mirrors or along the dashboards of their cars]. I know it sounds strange, but right now, with everything that's going on, I just feel safer, . . . more comfortable. I feel like this person is a part of my community; it makes me feel safer."[24] Nicole's words spoke to the fact that in getting around a Beirut under stress, certain kinds of sectarian seams, which are deeply enmeshed and play out in politics, often become thicker.

One afternoon in summer 2010, I visited the offices of an NGO and, as is not an unusual occurrence during the research process, the person I had arranged to meet was not there. Another member of the staff, Nabil, who looked to be in his early forties, brought me to a small conference room and shared some information and literature about the organization. Our conversation turned at some point away from the organization itself as he began to volunteer his thoughts about the current political situation in Lebanon. As a foreigner, I was often told, even in casual conversations with taxi drivers and shopkeepers, that I didn't understand the problems in Lebanon. Often, after I was told that I didn't understand how Lebanese politics worked, strangers would inform me by talking about corruption and Lebanon's status as a pawn in the geopolitics of the region. In this conversation, however, talk of Lebanese politics and society took a familial turn.

The problem in Lebanon, Nabil started out, "is that we do all this labeling. We want to figure out what someone's background is; we do this even from their name. Okay, this is your first name, that might tell me something about you; now I want to know your family name so that I can find out more." Indeed, I had observed how some Lebanese posed a series of questions when meeting someone for the first time, questions that took the form of assembling clues—such as first name, family name, neighborhood residence, father's village—that might help to sort out an individual's sectarian and sometimes class (through family name and residence) background.[25] I also met Lebanese who did not engage in this line of questioning and demurred when the questions were asked of them. One middle-aged professor had told me that some people thought that these inquiries, by differentiating people into sectarian types, were one of the everyday and seemingly banal—but extraordinarily powerful—ways of giving support to the broader structures of political sectarianism.[26] Ramzi, the professor, had said, "I refuse to play this game of talking about where my parents and family are from so that people can try and figure out what sect I belong to. . . . In my view, I do not belong to any sect. This is a game we play that keeps sectarianism alive and well."

According to Nabil, however, this mode of "labeling," as he termed it, was a way of determining whom you can trust. Essentially, he said, "it's a form of survival." But, in speaking about his two "tween"-aged sons, Nabil was more sanguine. As he described it, the globalization of media and the internet, as well as the pull of global consumer culture, shaped not only the activities and aspirations of his children but also their raison d'être in ways that seemed to bring to the fore generational differences among younger and older Lebanese. As one example, he said, his sons were far less likely than older generations had been to take up arms and fight to defend their sectarian community: "My sons see everything on TV and on the internet, they have their basketball, their rap music; they want to go out with girls and buy the newest car. . . . The men my father's age, when they were young, they didn't have anything going on in their lives. . . . When the leader came and said, 'Join us and fight,' they did. But my sons, the young people now, if war broke out and the leader came and said, 'Join us,' they'd say no; they're more interested in the things they want to buy, their music; these are their concerns."

Nabil's comments were reminiscent of the finding in a UNICEF-sponsored report that Lebanese youth "were influenced by the rise of global youth culture and enriched by the intensive global communication networks and the mass media at the same time that they felt stuck in parochial practices that entrap them in confined milieus" (Khouri 2011, 14).[27] In a conversation with Dania, a project officer for a civil-education NGO, she described these "parochial milieus" in much blunter terms: "I have a three-year old, and I want to keep him away from all of it [Lebanese party politics]. They are all liars. In my job, what we want to do is to create an active citizen instead of one who blindly follows the leaders. For me, as a mother, ... I don't want my son to go with any party."

But how does one reside on the outside of politics in a place divided by both a past war and one people anxiously feel is yet to come? When speaking about what it means to live in such a contentious political environment, which gives rise to occurrences of political violence and creates feelings of anxiety about the possibility of large-scale war, Nabil went on to draw me a rudimentary map of Beirut that outlined some of its neighborhoods. The safest thing for people to do is to avoid living along the borders of the different neighborhoods, he said, because of the risks. "You have to live in the heart of the territory because, in times of war, you need to be in the middle surrounded by the same community. That's where we live, in the center," he said as he marked the spot on the map that represented his neighborhood in a suburb of Beirut. He continued, "It's just safer this way." His point recalled the ways in which, for residents of places mapped by conflict, knowledge of the geography not only is a way of living a normal life but can also be understood as a matter of physical safety.

Nabil's remarks about the dangers of the city's political sectarian geography and his vision of Lebanese youth as being captivated by consumerist desires and global forms of media and entertainment were also striking on several other levels. First, his words told of how these seemingly inconsistent or incompatible activities, of global consumerism on the one hand and the machinations of political sectarianism on the other, might come together, in fact, to shape the everyday lives and mobilities of young people in Beirut. Although his sons may be absorbed with video games and rap music, their lives are also shaped by what their father described as the need to protect themselves in a place where the possibility of the eruption of armed political sectarian conflict and urban warfare is very present.

Second, Nabil's perspective about safety being constituted among one's own community was reminiscent of the way Nicole described her sense of security, which came, during times of heightened political tension, from riding in taxis driven by drivers she knew were from the same religious background as she. Both Nabil and Nicole described how shared communal affiliation was understood to be a site of shelter.

Finally, Nabil's narrative about his sons' consumerist-driven and mediacentric lives resonated with one of my own findings about the youth I spoke with who traveled around the city in order to pursue consumption-based leisure experiences. For many of these young men and women, the city was open for their exploration, a fact that spoke not only to their stations of privilege but also perhaps to their generational inclination to be less concerned with—or constrained by—the boundaries between people and place that were produced by the experience of living through the long civil and regional war. For Layla, the mother who wanted her two teenagers to stick close to home and hang out at a shopping mall in their neighborhood rather than one across town that she thought was a potential bombing site in 2005, the war had taken a clear toll on her geographic mobility. "I first started driving when I was twenty," she told me, "and it was during the war. It was a time when I was so stressed out and my reflexes were terrible. After I got into a small accident, I just decided I couldn't take it. I gave up driving then. I really do blame the civil war for the fact that I do not know how to drive!" When I spoke with her, she was driven around primarily by a hired driver whom her family also employed as an assistant for their business. Layla had thus established patterns of work and family life that restricted her daily mobility to just a few neighborhoods on the western side of Beirut, and she wondered aloud why her kids needed to go to different parts of the city.

In one sense, Layla's comments reminded me of the archetypal parent-teenager relationship, with the teenager straining against limits set by the parent. Yet, in another way, her children were experiencing life in a city where political sectarianism was very much still a geographical force but also one increasingly marked by the apparatus of consumer capitalism. In Layla's view, her children were caught up primarily with the process of consumer capitalist culture and showed less concern about the political sectarian mappings of the city—and their relationship to potential violence—than she did. In short, given her experience of living through Lebanon's long war, the city was mapped differently for Layla than it was for her children.

Looking closely at the mobility practices of youth in Beirut offers a window into important aspects of both political sectarianism and class, key categories of social difference in Lebanon. What these youth mobilities make visible is a complex set of social and spatial relations, characterized by the privileging of class and cultural positioning, that cross politics and sect during times of low political tension, on the one hand, and a kind of shoring up of political sectarian solidarity that is manifest during periods of heightened political tension on the other.

DIVERGING ROADS

In March 2005, amid the heated contention between the just-formed March 8th and March 14th political camps, I took a taxi out toward the airport to see traffic signs that had appeared overnight along the median of the highway. They were, in fact, faux traffic signs that used the commonly recognized symbols and vocabulary of traffic control to express political sentiments

FIGURE 3.2. Traffic sign: "No foreign intervention! 1559 prohibited from passing." (Photo by author.)

against U.N. resolution 1559, which called for the complete physical and political withdrawal of Syria from Lebanon and criticized foreign (read: U.S., Western) intervention in Lebanese political affairs.

While no group took credit for installing the signs along the highway median, their viewpoint belonged to the March 8th coalition of Hizbullah and its allies. Beyond the challenge of crossing the busy highway on foot to reach them, what struck me about the signs was how they illustrated the ways in which the theater of politics plays out in urban space. If the city, as theorist of space Henri Lefebvre conceived it, is a "place of encounter" ([1968] 1996, 158), then in Beirut even a routine trip to the airport can entail a spatial confrontation with the fervent messages and symbols of sectarian-based party politics. Moreover, while political polarization is not a phenomenon unique to the space of Beirut, the ways in which politico-ideological issues spill over into space and are incited by domestic and regional geopolitics—like the passage of U.N. Resolution 1559, for instance—gives everyday use of and movement through the city particular meaning. While Beirut is no longer precisely divided into warring fronts, as it was during the protracted civil and regional war, the mobility of the city's residents is still shaped by—and also shapes—Beirut's divided and sometimes volatile political sectarian geography.

In this chapter, I hope to have accomplished several things. First, I have tried to show that political sectarianism is an enterprise that takes form in urban space in various ways, one of them being the marking of territories through visual means such as banners, flags, and graffiti, as well as through the appearance of political violence. Second, while class and status are often understudied in analyses of political sectarianism in Lebanon, I consider what the urban mobilities of privileged youth show us about the complex relations of class, politics, and sect that characterize everyday life in Beirut. In so doing, I have sought to highlight some of the ways in which residents' uses and understandings of urban space play a critical role in producing the city's public and political landscape.

4 · SECURING BEIRUT

I heard the first bomb go off on October 1, 2004. I had arrived just weeks before to begin research and was staying at a residential hotel when I heard the blast—a blast I thought was coming from one of the many construction sites near the downtown waterfront area. It was in fact the explosion of Marwan Hamadeh's car on a sunny Friday morning. He was being driven along the seaside road in Ain el Mreisse, a central neighborhood adjacent to downtown, when a bomb ripped through the front of the vehicle. His driver, Sgt. Ghazi Abu Karroum, was killed instantly and Hamadeh and his bodyguard survived with injuries. Like other political figures with ties to prominent Lebanese sectarian-identified kin groups, Hamadeh, the outgoing minister of economy and trade and future minister of telecommunications, had cycled through a number of powerful government positions during his career. In the months prior to this attempt on his life, Hamadeh had been vocal in his opposition to the Syrian regime's efforts to amend the Lebanese constitution in order to extend the Lebanese presidential term and, in so doing, lengthen the presidency of Emile Lahoud. Lahoud was thought by many to be a Lebanese figurehead for the Syrian regime.

The attempt to assassinate Hamadeh marked the return of violence to the streets of Beirut more than thirteen years after the end of the civil war. This return was announced by the horrific spectacle of exploding bombs that killed political figures and journalists when they entered their cars and turned the ignition key, or, later, when they were driving or being driven in the city. After the bombs went off, various kinds of security configurations would be set up throughout the city, but particularly near the homes, workplaces, and leisure areas of political VIPs. As a result of this intensification of security to protect political figures and public buildings, the city became increasingly militarized, and getting around involved particular kinds of spatial and social negotiation.[1] As elsewhere in the world, processes of securing the city involved the reconfiguration and management of everyday mobility in public space. Efforts to organize and control movement in Beirut produced, as they have in the post-9/11 United States, for example, certain types of insecurity engendered by fear but also, importantly, by the conditions of security themselves. During the 2004–2006 period, amid the series of bombings in Beirut, most of which were targeted assassinations, mobility was regulated by a security apparatus comprised of both privately hired security personnel and the state's police and armed forces, who created lines of defense intended to protect particular people and places from harm. Residences, luxury hotels, offices, and governmental structures in which an overlapping group of class and political elites lived, worked, and played were the main sites of securitization.

One type of use of public space in particular was understood to be most suspect and to pose the greatest threat to forces of security: driving. During this time, as during the protracted civil and regional war (1975–1990), the vehicular bomb was the signature means of violence: in almost all cases, individuals were killed or injured while driving, being driven, or passing by cars. With every car imagined as a possible bomb, "the rules of the road changed" (Packer 2006, 379).[2]

In this chapter, I explore how processes of securitization shaped daily experiences of getting around Beirut during the 2004–2006 period. Security measures transformed not just the city's built environment but also the mobility practices of its residents, as motor vehicles and, in some sites, pedestrians were deemed a potential threat to a select group of people, namely political figures, while they were at home, work, leisure, or in transit themselves. Focusing on the ways in which mobility was circumscribed through the setting up of barriers, blockades, and checkpoints and the rerouting of traffic flow, I explore how security installations adopted mainly by and for the power holders created conditions of unevenness in the urban landscape. In seeking to effect order in and around certain urban spaces, security made getting around the city disorderly. My study of security builds on explorations of how security is spatialized in the urban environment. Scholars have investigated a diverse range of topics that highlight the importance of urban space as a locus, medium, and tool of security strategies: the global ascendance of gated and privately secured forms of residence; the role of surveillance in the fragmentation of the urban environment and the rise of private security forces; and the emergence of vigilante groups that provide extrastate security to protect citizens from crime and violence.[3]

The fact that security is oftentimes a joint private and state enterprise in Beirut, a phenomenon that has been referred to as "multiagency policing" or "plural policing," is not unique.[4] However, the Beirut case appears distinctive because of the extent to which the city's "securocracy"[5] spatially represents Lebanon's multiple sovereignties of state and nonstate actors—or what geographer Sara Fregonese (2012) refers to as a "hybrid" sovereignty—rather than one coherent state sovereignty. In other words, the multiple secured areas around which city residents have to navigate, where privately employed bodyguards, domestic police forces, and national army soldiers work alongside one another, exemplify not just one of the means by which public space is increasingly being privatized, a topic I discuss in chapter 1, but also the blurring of the lines between the state and the nonstate that are an important part of the makeup of Lebanon's physical landscape.

THE ZIG-ZAG AND OTHER ASSORTED BARRIERS

On an evening in February 2006, one year after Hariri's assassination, I sat in a taxi heading from Verdun, a neighborhood in central Beirut, as the driver slowly weaved back and forth in a zig-zag motion around a series of metal barriers. Set up for a distance of about 20 or 30 meters, the barriers were a temporary and sudden obstacle to be overcome by drivers. To keep from crashing into them, the taxi driver had to adopt a back-and-forth motion, turning the steering wheel to the left and then to the right. When I asked the driver why the barriers were there, in this section of the road only, he said it was because of security, and he pointed out that we were driving past a police station.

Barriers set up in a zig-zag pattern were among the types of security installations suddenly and temporarily erected in response to a sense of heightened vulnerability surrounding particular places and people. In spring 2005, as I described in the previous chapter, a pattern seemed to be forming amid the spate of bombings that followed the killing of former prime minister Hariri: predominantly Christian parts of Beirut were repeatedly the site of bomb blasts, and high-profile journalists and political figures associated with the anti-Syrian March 14th alliance were targets for assassination. As a result of the emerging pattern of attacks, security was thus built up throughout the city at locations that bore affiliation—actual or perceived—with the politics of the March 14th group.

On Good Friday in March 2005, for example, a few blocks from my apartment building in Koreitem, I came upon yellow police tape surrounding a Catholic church. Extending for about 30 meters along the sidewalk directly in front of and across from the church, the tape, and the soldiers with machine guns positioned adjacent to it, obliged pedestrians to walk for a ways in the street. Like other security installations, this one around the church was many-layered: the yellow police tape, though made of flimsy plastic, clearly demarcated areas of restricted access, and this cordoning off of public space was buttressed by armed soldiers with machine guns at the ready. This security set-up was in position only for the duration of the Easter weekend, a time when the perceived threat of attacks on Christian-associated sites in any part of the city was intensified given the symbolic value of the holiday and the large numbers of attendees at church services.

Yellow police tape also appeared in spring 2005 to cordon off the front of the luxury seaside Mövenpick hotel. At the hotel, the rich and powerful both worked—it was regularly the site of meetings of political dignitaries as well as the meeting place for U.N. investigators looking into Hariri's death—and played, as it was equipped with one indoor and three outdoor swimming pools and several tennis courts, among other amenities. Because of its use by political elites, both domestic and foreign, the hotel was considered a potential target for violence. The barrier created by the yellow police tape went over the sidewalk and extended into the busy seaside boulevard that passes by the hotel. Thus, in going around the barrier, pedestrians had to leave the sidewalk as they approached the hotel and walk in the street alongside speeding cars. The tape was not a temporary fixture, as the one bordering the church had been; it remained for more than a year. A staff of armed security guards near the hotel's vehicle entryway formed a second line of defense, just behind the yellow police tape, and their presence ensured that walkers accepted the limitation imposed by the barrier by exiting the sidewalk and continuing in the street.

Security installations differed not only in their material composition but also in their temporal and spatial dimensions. While some formations of security, like the metal barriers set up in a zig-zag pattern and the deployment of soldiers at the church on Easter weekend, were enacted on a short-term basis and could be easily dismantled, other formations were semi-permanent and became more fixed into the built fabric of the city; they overlaid public space with a political gravity, as Ochs (2011) remarks about the context of everyday security in Israel. Because parked cars were the most frequent vessels for bombs, the prevention of vehicular parking was a chief aim of security efforts, and barriers to parking appeared in various forms. Steel bars soldered together into what resembled enormous and fearsome toy jacks were set up along the roadside adjacent to a building that housed the offices of the Future Movement, the party led by the Hariri family and a leading member of the March 14th alliance. What looked like painted concrete shark fins fixed onto the street provided another type of barrier that restricted parking alongside edifices. In the downtown area that was home to both high-end retail stores as well as the parliamentary offices parking regulations that had gone unenforced were suddenly given attention, according to Elie Helou, a transportation project manager for the state-affiliated Council for Development and Reconstruction. He described how "what often looks like traffic enforcement is actually security.... Before February 14 [the date of former prime minister Hariri's assassination], there was lots of parking on that main street downtown, Bank Street. And now the police are sweeping the street and enforcing the no-parking rules on that street, but this is because of security issues, not because of an interest in managing the traffic situation overall." In this way, security became an integral, even prioritized, aspect of broader urban planning processes that shaped how the city's public space would be used and managed.

Still other security formations might be best described as neither temporary nor permanent but, rather, itinerant. A regular sight in public life are caravans of security personnel accompanying political VIPs en route for business or pleasure. These caravans use vehicular and verbal means to suddenly storm and clear streets in order to allow for their breakneck and unhindered passage. In their harnessing of mobility itself as an instrument of power, these vehicular convoys demonstrate both the portability of and

FIGURE 4.1. Shark-fin barriers. (Photo by author).

the impromptu aspects of security. Typically consisting of black SUVs with blacked-out windows that often bear no license plates, the convoys are only sometimes led by a police vehicle with flashing lights or a siren. More often, they appear without warning to take over the field of drivers. Thundering past motorists frantically trying to get out of their way, the convoys activate state power and its roving security project in public space through their mode of mobility but also, in their towering height, bulk, and "unmarkedness," these vehicles go beyond the bounds of the state to represent a more generalized form of elite untouchability.

Once, on the highway just outside of Beirut, men in vehicles that were part of a convoy moving at breakneck speed leaned out of the windows of their SUVs gesturing and yelling at the driver of the car in which I was traveling. They appeared incensed that she had not moved from their path fast enough, although their cars had surged from behind at such a rapid speed that there had been little time to react at all. A second later, a car hastening to move off to the side as the convoy came upon it crashed into the highway barrier. We caught just a glimpse, but as the convoy continued on, we were able to see that while the front of the car was badly damaged, the driver was thankfully unhurt.

The crash illustrated the ways in which vehicular behavior on the part of agents of the state and its proxies produced spatial—and uneven—risks in the public environment. While acting to secure public space to allow for the swift and unconstrained movement of the powerful, the convoys and the "extrahard exoskeletons" (Miller 2007) of their SUVs create physical dangers for other drivers on the road. At the same time, by making the movement of ordinary users of the street disorderly and secondary to that of the VIPs and their need for a secured and mobile buffer zone, the convoys engender a kind of class injury through the means of automobility.[6] As a floating form of security that glides over the city, the convoys are part of the broader private takeover of public space that has characterized Beirut's urban development in the rebuilding era following the protracted civil and regional war. Unlike the barriers, which prevent mobility by stopping, halting, and prohibiting, vehicular convoys place mobility at the service of security in order to allow elites to live a cocooned existence that takes them, by means of protective SUV capsules, from one secured urban space to another.

PRIVATE-PUBLIC SECURITY

While the manipulation of the urban built environment as a means of regulating mobility and as a technique of crime prevention is not a new phenomenon, scholars of urban space have tracked its intensification around the globe in the late twentieth and early twenty-first centuries. One area of this scholarship examines how middle-class fears about and perceptions of the violence and crime that dwell in urban areas have given rise to new forms for securing domestic and residential realms. By fortifying their residences and neighborhoods with features such as surveillance technologies, gates, walls, and floodlights, property owners seek to create a community that "includes protecting children and keeping out crime and others while at the same time controlling the environment and the quality of services" (Low 2004, 230).

These mechanisms of homegrown security also mark certain public spaces—the streets that border or run through residential areas, for instance—as the province not only of the homeowners and their guests but of a raced and classed set of "insiders" who do not arouse suspicion. "Outsiders" are unwelcome and are often subject to scrutiny or worse. Secured spaces, intended to protect their denizens, thus operate with a racial, class, and gender logic that works from ideas about who belongs and who does not, who is threatening and who is not. In parts of the suburban United States, for example, private home and commercial-property owners seeking to protect and secure their surroundings and assets work in collaboration with community policing and neighborhood-watch groups to report suspicious activity to law-enforcement agencies. Security is enforced, on the basis of this logic, with sometimes deadly consequences, as armed residents aim to weed out threatening "others" from their terrain.

In Beirut, this kind of residential and neighborhood-based security, which draws on and collaborates with state-level policing agencies, was also in evidence. Although during the 2004–2006 period security installations surrounded the sites of work, leisure, and residence of high-profile individuals and thereby represented modes of privatized security being mapped onto a public landscape used by all, these individuals were protected by privately hired security guards as well as by members of the state's police and armed forces. What is important about this fact, that security was an endeavor that joined together staffs of private bodyguards with state forces of policemen and soldiers, is the ways in which it made visible in the city's space the blurred lines between state and privatized provisions of security. The intersections of public and private security, themselves rooted in the overlaps between Lebanon's class and political elites, brought into sharp relief the kinds of challenges to notions of the common good that inhere in Beirut's civic space. In other words, the activities of the ordinary citizen in urban life were in a precarious position vis-à-vis the public-private configurations of security that transformed the rules and routes of spatial movement. These enactments of state and class power, moreover, through their imbrication of public and private security, engendered a particular politics of social and spatial exclusion.

SECURITY PROFILING

Nearly all residents engaged in modes of spatial and social negotiation in order to navigate the barriers and blockades that regulated the vehicular and pedestrian movement of most—while at the same time making openings for the movement of a select group of others—but dimensions of inequality shaped the dynamics of these negotiations as certain kinds of residents were subject to a heightened degree of scrutiny as they moved through the city.

During the period of political crisis and violence in 2004–2006, opposition to the Syrian administration and the presence of its army in Lebanon, a long-standing sentiment that had endured since the time of the long war at the onset of the Syrian occupation, reached fever pitch as a result of widely held beliefs that Syria was behind Hariri's assassination. This opposition to the Syrian regime, which constituted the basis of the platform of the March 14th alliance, spilled over into attacks on Syrians themselves. As a visible underclass linked with what some Lebanese understood to be an enemy state Syrian workers constituted a socially vulnerable—and visible—group moving through public space. Construed as potentially threatening in an anxious political climate, as well as powerless, they were uniquely regulated by security forces.

"Even if I am *nizami* [law-abiding], I get pulled over by the police," one Syrian laborer who worked delivering fast food by motor scooter told me. Even though in possession of updated driver's licenses and properly maintained vehicles, Syrian delivery workers I spoke with described being detained at roadsides, at police stations, and having their vehicles frequently seized. One restaurant worker thought that Syrian delivery drivers on motor scooters were most suspect because he had heard rumors that "they [security forces] think that it might be a scooter loaded with explosives that is setting off some of the bombs." "But they're not sure," he added. While the mobility of Syrian workers was subject to an increased level of scrutiny by the security apparatus even prior to Hariri's assassination because of their status as a disenfranchised population, the political tone and precariousness of the post-Hariri period exacerbated their vulnerability. Indeed, the bodies of the workers were profiled by Lebanese security services in ways that bore similarities with U.S. security agencies' profiling of African Americans, Middle Easterners, and South Asians engaged in practices of mobility. As part of the U.S. vernacular, tongue-in-cheek expressions like "driving while black" and "flying while Arab or Muslim" (in the post-9/11 era) reference the policing that is generated by these profiling practices. These kinds of practices, like those the Syrian workers encountered, are initiated by suspicion of certain raced and classed bodies. In this way, while the order implemented by security in Beirut disrupted mobility for many, it often rendered others, like the Syrian workers on motor scooters, immobile.

In techniques of securitization, risk is anticipated not only on the basis of ideas about race, gender, class but also through biometric technologies of profiling that purport to identify and detect likely, future, offenders through, for example, reading and analyzing minute aspects of the body such as facial expressions that betray criminal intent. Studies of the global regime of mobility that consider technologies of security in the post-9/11 era highlight the use of biopolitics, which examines and interprets bodily gestures in seeking to surveil, to assess the risk presented by, and to manage the movement of individuals. These programs of behavioral profiling, unlike the zigzag barrier or steel blockade, bear a futurist temporality that creates what geographer Ben Anderson calls "geographies of anticipatory action" (2010). By acting upon individuals "on the basis of behavioural potentialities rather than on the basis of how they have actually acted" (Bell 2006, 160), these technologies of security bear a distinctively preemptive character.

In Beirut, security also acts on the basis of behavioral potentialities, but in a way that makes central social standing as opposed to affective and bodily gestures. For example, the barriers that require the zig-zag movement are often set up this way in order to slow cars and allow guards to take a longer, more substantive look at a vehicle's occupants. Suspicious drivers are told to pull over and thus enter a second, verbal and more discerning, stage of inspection. Though the criteria for and assessment of who and when someone is deemed suspicious is contextual, it draws significantly on the inextricably linked dimensions of social status and potential criminality rather than on the kinds of technologically sophisticated forms of spatial surveillance employed in many other sites around the world.

In Beirut's public space, a labor force of armed guards typically measured people's behavioral potentialities. Members of security, whom I observed to be exclusively male, seemed to identify and detect the level of threat posed by an individual or group according to both broad and localized notions of social hierarchy and identification that ranked and sorted persons by making use of various sensorial registers including the visual discernment, verbal communication, and mental determination of characteristics such as class, gender, age, and religious and national background. Arabic accents, the wearing of religiously identified clothing such as the headscarf, and one's status as a Western foreigner are all features of the complex matrix that is the basis for the calculation of threat level.[7] Indeed, as Fawaz, Harb, and Gharbieh observe about threat detection in Beirut, even the wearing of plastic shoes like flip-flops, which are thought by many Lebanese to be inferior-status footwear appropriate only for the interior of the home, might give security personnel enough reason to conduct a more thorough going-over (2012, 182).

Sara, a mother of two in her thirties, related to me how, in her verbal interactions with members of Beirut's security forces while driving (a Volkswagen), she always tells them that she is in a rush and "needs to get through" (if there is a barrier or blockade) in order to pick up her children at their friends' home. She recalled how this tactic of emphasizing her role as a mother trying to reach her child often proves successful at getting around the constraints security might otherwise place on her mobility. Sara's identity as a woman, a mother, and upper income (driving a stylish German car) is recognized by the localized regime of mobility administered by the security forces as one that is entitled to greater freedom of movement. In this sense, amid an urban social space structured by patriarchy, understandings of gender and class come together to create a kind of mobility of entitlement whereby the upper-class mother is configured as both powerful and, in a certain sense, powerless by the first line of security. In negotiating for access, Sara was able to draw on a form of power that stems from her elevated class position and, at the same time, gendered notions of women and mothers as unthreatening and in need of male guidance and protection. Her access to urban space stands in stark contrast to the experiences of the Syrian worker on the motor scooter, who held low status in Beirut society and was deemed possibly threatening by certain political groups. His movement was circumscribed by security. This inequality in access produced a stratified mobility whereby movement through the city by some urban residents, like the upper-income woman in transit to pick up her child, was less suspect and more unrestrained than that of others—the male foreign laborer, for instance.

NAVIGATING SECURITY: TACTICS, STRATEGIES, AND COMPETENCE

In Beirut's realms of security, social class, national origin, and gender are important facets of the complex social matrix upon which decisions about spatial access are made. However, these features of social identity are not the only factors that go into determining to whom spatial access should or should not be ceded. One's ability to negotiate with security forces can also shape the outcome of one's experience in and around a secure area. The security encounter, like other interactions in Beirut's public life, might be navigated through familiarity or by successfully mobilizing one's social capital through perceived, proclaimed, or demonstrable affiliation with well-placed individuals, often kin members, or institutions.

Once, while riding as a passenger in Maya's car while she ran errands around the city, she entered a parking lot that sat across from the building housing a prominent NGO. She had picked up and dropped off documents at this building many times, and the men who staffed the parking-lot security gate knew her, she said. This time, the security guard chatted with her and told her that although he was supposed to check under her car—using the wheeled search mirror he was holding—he was not going to bother her with this check, and he gave her clearance to go ahead and park.

Reminiscent of Erving Goffman's dramaturgical perspective of urban public life as an arena in which individuals present their "familiar" and "connected" selves (Hannerz 1980), Maya's mode of gaining spatial access in a city characterized by various forms of security highlights the significance of face-to-face interactions in Beirut's urban environment and challenges assumptions about cities as spaces with high levels of both anonymity and anomie. Moreover, the cultivation of the "familiar" and "connected" self highlights the harnessing of certain sets of skills and tactics that residents use in getting around the city.[8] If, as anthropologist Ulf Hannerz (1980) notes, we conceive of the "city as theater," the security encounter is a scene through which residents as actors display, or hope to display, a kind of "user competence" (Fawaz, Gharbieh, and Harb 2009, 2) that enables them to navigate past social and material security barriers. The privileged, in possession of a greater level of user competence, often, but not always, find a means of access.

In February 2006, I escorted a group of students, who happened to all be women, from a colleague's anthropology course at the American University of Beirut (AUB) onto the busy commercial street adjacent to the university to participate in a classroom exercise about taking fieldnotes. The university's picturesque and gated campus, replete with red-tiled roofed buildings boasting Mediterranean views, is an oasis of green amid the concrete of Beirut. With its English-language curriculum and costly tuition, AUB enrolls students primarily from privileged backgrounds.

It was a rainy morning, and although we had planned to focus on completing the classroom exercise, matters of security prevailed. We exited the campus, heading past one of the several entrances around the campus's perimeter staffed by guards, and stood under the awning of a restaurant to take notes. This section of the street was dotted with soldiers, and in front of them cars moved along the one-way corridor zigzagging back and forth between a corral of metal barricades. The then-prime minister, Fouad Senioura, lived in an apartment building on this block. Passing along in front of his building was a meandering, vehicular caravan overseen by a group of soldiers. A minute or so after the students began taking notes, a soldier came to ask us what we were doing. The students took the lead, responding that they were from the university and were doing an assignment for class. The soldier left, and we continued observing and writing. A few minutes later, another soldier, this one more imperious, approached our group. "Could you move?" he asked rhetorically. Flustered, we made the collective decision to find another location to complete the exercise. In their fieldnotes, the students wrote about what happened. One wrote: "What I was thinking about during this conversation [with the soldier] is that, are we a kind of threat to them? Are we a threat to security? Aren't the streets for the public? After this experience with the solider, I felt uncomfortable to stay there; we were violating the rules of the public space that is now part of several private spaces: the restaurant and Senioura's property."

Other students also commented that the soldiers' telling us to move demonstrated the insidious expansion of the private at the expense of the public. Their notes voiced not only resentment about this takeover of public space but also uneasiness about having their freedom of movement curtailed by the soldiers, in particular, I surmised, because they were unaccustomed to being viewed or treated as threatening in public space. One student, for instance, spoke in her write-up about how the stares of both the soldiers and passersby made her feel as if she was doing something wrong. "Were we a threat?" she wrote, "I hope not. I don't want to get arrested just for writing notes."

I shared the students' sentiments insofar as I too found the encounter both surprising and unsettling. Having come and gone from the university on a regular basis, I was, as I am sure the students were, aware of the increased security along the block where Senioura lived, but I presumed, mistakenly, that our behavior on the street would be perceived as harmless. What is more, I thought that our association with the university, a province of the privileged, would grant us access to the street that others might not be permitted. I thought this association furnished us with the social capital that would enable us to move freely in the public-cum-privatized space, as the one student summed it up, that this block of the street had become.

But security encounters are variable and, as such, one can predict neither the course of action of an encounter nor its outcome. For example, it might have been our act of writing notes that bothered the soldiers, just as the taking of photos on a public street near a blockade—or even the carrying of a camera—might incite a confrontation with security personnel. But we also might have tried to negotiate with the soldiers by claiming that we had permission from the university administration to conduct a classroom exercise and to record information there on the street. This claim of authorization from high-ranking authorities, a demonstration of our competence as users of Beirut's urban space, could have thrown off the soldiers and entrenched our position. However, in the context of security, mobility and access are contingent on a myriad of factors including, but not limited to, the moods and whims of the soldiers and guards, the time of day, and the prevailing scale and dimensions of political tension.[9] In Beirut, there are no state-issued red or yellow alerts indicating the level of threat,[10] but the ebbs and flows of political crisis, which in turn shape security measures, play a key role in the management and regulation of spatial movement.

Moreover, what is particularly interesting about the nature of Beirut's installations of security during the 2004–2006 period is the way in which they demonstrate the hegemony of party politics over the interests of the public or common good. As urban studies scholars Fawaz, Harb, and Gharbieh note (2012, 188), security deployments in Beirut show not only disregard for the common good, as the comings and goings of a select group of influential individuals curtail and alter the movement of the majority of the city's residents in various ways, "but actually the absence of a claim forwarded in the name of a 'common good' to justify these deployments." In fact, it would seem that the securitization of key political figures and efforts to maintain their unhindered mobility through the city to home, work, and leisure outweigh not only the public interest but also what Michel Foucault understood to be the key aspect of the city's spatial order: making possible, guaranteeing, and ensuring the circulation of people, goods, and money (2007, 29). The clout of those whom security surrounds is such that their circulation across "archipelagos of privilege and power" (Graham 2010, 96) is given priority over the routinized activities of public and commercial life. The workings of security, in this way, have an exclusionary effect in the urban environment.

ENCLAVIZATION

Security reconfigured mobility in Beirut in both the short term, as weaving back and forth between barriers was suddenly required, and also in more lasting ways, as residents had to find new routes from home to work or from school to home because of road closures and the rerouting of traffic flow in areas near VIP residences.

Among the most heavily fortified spaces in Beirut after February 2005 was that surrounding the home of former prime minister Hariri, a palatial block-long structure in the west Beirut neighborhood of Koreitem. Following Hariri's assassination, the street passing in front of the home was turned into a one-way thoroughfare with checkpoints at both ends; security was later intensified, and vehicular access was denied entirely. Here, both personnel employed by the Hariri family and Lebanese armed forces managed security.

The order enforced by the security near the Hariri palace, and elsewhere in the city, created a kind of disorder for many residents, especially drivers. On many of the occasions when I met with Maya, a professional whose workplace was adjacent to the Hariri palace, or rode as a passenger in her car, she complained about how the street's being changed to a one-way route added time and stress to her commute. In fact, not only was traffic flow on the main thoroughfare that passed by the palace disrupted, but also circulation along the smaller, connecting streets was managed by guards and signs announcing that the streets were closed except to those who needed to conduct business at the palace. Also, sections of a busy road down from and parallel to the one in front of the Hariri palace were lined with yellow police tape that read "police line" in English and displayed numerous "no parking" signs.

Maya's everyday driving patterns were caught up in the development of security enclaves throughout Beirut. Similar processes of enclavization have shaped the everyday lives of residents in other cities in the Middle East. In Cairo, for example, exurban gated communities have been expanded, as have urban elite consumer practices that involve patterns of travel from one type of guarded space to another, a phenomenon geographer Stephen Graham has termed "passage point urbanism" (2010). Elsewhere in the world, fear of crime has led middle- and upper-class residents, in São Paulo for example, to fortify their residences and neighborhoods, employ private security forces, and seek out modes of urban mobility that enable them to pass over, in some cases quite literally through the use of helicopters, the urban streetscape and the likelihood of interaction with its lowest-status denizens (Caldeira 2000).[11] These kinds of classed processes of spatial and social segregation link up with the broader global phenomenon of establishing secure, private, separately administered, and "globally networked" territories within parts of nations (Ferguson 2005).

Navigations of the territories of security in Beirut took place in a public space already configured by relations of power wherein certain bodies are subject to increased scrutiny and regulation. In processes of securing Beirut, these inequalities of spatial access and mobility were produced in concert with forms of symbolic domination experienced by all. For example, following his assassination, photo billboards of Hariri were put up downtown, in West Beirut, and along major highway corridors surrounding the city. His oversize image, draped from verandas of buildings facing one another, hung over the center of narrow streets in neighborhoods affiliated with his political party and its supporters. A graphic production company with whom Maya dealt for her work told her of the huge profits they had made printing Hariri posters. It was a continuous source of revenue, they said, because the poster images were replaced with new ones every few months. Other Hariri displays appeared in electronic form; a digital scoreboard adjacent to his downtown burial site and another near the studios of the television network his family owned kept a running count of the number of days since his death. As a political technology visually representing and memorializing Hariri's power and surveilling public life, these pictorial and numeric displays constituted territories of security in their own right.

Security enclavization was also a feature of the neighborhood where the speaker of the house of parliament, Nabih Berri, one of the top three men in Lebanon's political system, resided. In Ain-al-Tiné, just west of Verdun, on the street where Berri lives, security spilled out into the lives of residents and passersby. Hania, a woman in her forties, lives with her family in a building down and across the street from Berri's residence. It was early 2006 when Hania told me how Berri's security had recently reduced the height of the speed bumps that were staggered along the street. After being installed in order to slow vehicle traffic from one end of the street to the other and thereby enhance surveillance capabilities, Berri's neighbors complained that the bumps were too high and caused damage to their cars. But soon, another type of barrier to mobility appeared: a blockade staffed by soldiers was situated at one end of the street. The other end of the street remained open, but midway up the street a gentler barricade stood: a low chain strung between two metal posts blocked cars from passing through.

Walking along this street, one closed to public vehicular traffic, created an intimate encounter with this barricade. Amid this intimacy, a consequence of the slower pace and bodily aspects of pedestrian mobility, the many layers of security were made evident. I met Hania at her building several times, and each time I approached the chain, because there were no sidewalks on her street, I thought about the ways in which the barrier might be breached: I could easily unhook the chain from the metal post or, more simply, step over it, for instance. But the chain represented only the first layer of security, as armed soldiers stood watch on the side of the street, often taking shelter from the sun by standing or sitting in a wooden sentry box. A cognitive and bodily awareness of this second layer of security made me automatically pause just in front of the chain and look over to receive a signal from the soldier(s) that I was permitted to proceed. Then, and only then, would I step over the chain. If I was not signaled and approved of from a distance, this seemingly penetrable barrier became even more cumbersome. Verbal communication added another layer to barriers, one of social negotiation and the increased possibility that the right of access would be denied.

For example, on one occasion, after passing the checkpoint at the opening of Hania's street, I sat on a ledge outside her building to wait for her. A soldier stationed in front of Berri's residence walked over and asked me why I was sitting there and whom I knew in the building. I gave him Hania's name. In an act that suggested the improvisatory, even arbitrary, nature of security, he proceeded to quiz me: "What is her family name?" "What floor does she live on?" It seemed that he knew all the residents in the building, even the floors on which they lived, and that I would be given permission to remain sitting there only if I could prove that I was indeed Hania's acquaintance. I erred a bit, giving Hania's parents' family name as I was unable to recall her married name, but the soldier was satisfied with my answers. He relaxed and began to make affable conversation.

At checkpoints and blockades not only social interactions but the material forms of the barriers themselves were unpredictable and required tools of negotiation. The changing and multilayered nature of the material barriers that manifested on Hania's street, for instance, exemplified the shape-shifting nature of security installations in Beirut: secured areas were increasingly and suddenly fortified as barriers took on harder, more permanent forms, bollards and no parking zones were erected overnight, and streets were unexpectedly closed off or were reconfigured, without warning, from two- to one-way traffic. In this way, processes of securitization made daily mobility a challenging and uncertain affair. "It's annoying," Hania said of the impediments that lined her street, but "it's better to have all of this than not to have it I guess. . . . Although it's interesting, because no one was killed in his house. All of them have been killed *a-tareeq* [on the road]." Indeed, those assassinated had all lost their lives in their cars.

In 2006, friends visiting the home of Karine, a twenty-one-year old college student, would encounter a checkpoint on arriving at her street.

Karine's street in the upmarket northern Beirut suburb of Rabieh was home to the residence of General Michel Aoun, a key figure during the Lebanese civil and regional war who reappeared on the political stage in spring 2005 after returning from living in exile in France. Security forces protecting Aoun's residence, which Karine thought consisted of both privately hired guards and state soldiers, issued visitor parking passes to households along the street for their guests. "I always tell my friends to call me when they get close to my house so that I can come out and give them a parking pass," Karine explained. What happens if you park your car on the street without a parking pass? I asked her. "They will tow your car," she replied.

Along Aoun's street, two distinct, yet intersecting, types of enclavization were at work. One type of enclave, as I discussed earlier, was established by mechanisms of security seeking to protect Aoun and his family from harm. Though not an officeholder, Aoun's personage generated processes of securitization that altered the mobility practices of his neighbors and their guests. However, the security enclave was enmeshed with another, that of the affluent. Situated en route to the mountains but only a twenty-minute drive from Beirut, Rabieh is an exclusive neighborhood of luxury hillside villas. In a sense, we can conceive of this intersection of enclaves as more than merely another example of the overlapping of social and physical geographies of class and political elites in Beirut. The intersection of enclaves changed things, if only a bit: wealthy residents accustomed to exercising their own forms of security and surveillance in the interest of maintaining not just safety but exclusivity found themselves subject to the scrutiny of bodyguards and soldiers. This change demonstrates not that the mobility of all residents was equally regulated and disciplined by Beirut's installations of security in the 2004–2006 period but rather the extent to which, in different ways and to different degrees, security shaped everyday spatial life for residents from many walks of life.

According to Maya, who had to reorder her route to work as a result of a security blockade near the Hariri palace, exercising control over traffic was an important way of inscribing power in the urban landscape, especially for the political elite. She complained, for example, that the Hariri family did not reopen the seaside road, a major traffic artery, near the site of the explosion that killed Hariri after the one-year anniversary of his assassination: "I mean, they could have reopened the road as part of the commemoration of his death in February 2006," she said. "I don't understand why they couldn't have done that, instead of just leaving it closed down and creating such a bad traffic situation for everyone, even after a year has passed. They decide for all of us how we are going to live in this city!" Here, Maya regarded mobility for all as coming under the control of a select few—Hariri's family members, to be precise. Her comment about how things worked in the city, about how the influence of the billionaire Hariri family extended even into the realm of traffic and road closures, resonated with everyday casual remarks I heard about how the "country is run by a few people." In this way, while the kinds of security assemblages that took shape in the 2004–2006 period could be viewed as an exceptional state of being, for many ordinary residents they represented business-as-usual politics writ large on the public landscape: the interests of the ruling class, to modify the classic line from Marx, were the ruling relations of space and society in Beirut, especially in the era following the long civil and regional war. Or as Nathalie, a university student, put it, "All the leaders have turned their neighborhoods into a *thakanat* [barracks]—you're going somewhere and then you find that a certain street is blocked and you can't think of a different route right away, . . . This is how it is; we're just forced to adapt."

SECURITY FOR WHOM?

In 2007, during the question-and-answer period following my research presentation at a workshop, a Lebanese woman criticized my analysis of experiences of security in Beirut. "You don't emphasize enough how security makes people feel safe," she said. She was right. I had not emphasized feelings of safety that security installations may have fostered for some residents. Thinking that I might have overlooked these sentiments about feeling safe and protected as a result of the security measures enacted in the 2004–2006 period, I went through my research data again.

During the 2004–2006 time frame, violence took the form mainly of a series of assassinations of political figures and journalists and several bomb explosions in largely uninhabited sites in predominantly Christian areas of Beirut. Later, in summer 2006, a conflict between Hizbullah and Israel escalated into full-scale war, and, in spring 2008, the deterioration of the domestic political situation brought the looming possibility of a return to civil war. Amid these two later contexts of crisis, the material formations of security in Beirut continued to surround high-profile locations, but the reach of violence expanded significantly.

But in my fieldnotes and interview transcripts from the 2004–2006 period, on which this chapter focuses, I found that residents expressed frustration, ambivalence, and weariness about the installations of security that took hold in the city rather than feelings of being protected by them. Nada, a young mother who lived in an apartment in Ain el Mreisse, only blocks away from the site of the explosions that took Hariri's life and from which Hamadeh narrowly escaped with his, spoke pointedly about politics, and security, in these terms: "The whole situation is very depressing. Everyone I know is depressed, when I call my friends we'll say that we feel tired, and we won't really say it, we won't say that we're depressed. But the political situation is making us depressed. So when you ask about security, all I can say is that it is for the big deal makers who decide the fate of the Lebanese. They will decide to either keep peace or to make war, and they. . . . Well, most of them have dual nationality—so if a real war does break out, they'll be the first to leave the country."

For Nada, mechanisms of security not only did not make her feel protected but were emblematic of the oligarchic dimensions of the Lebanese political system, in which a group of elites dominate the realms of government and play games of politics whose greatest risks—including the threat of full-scale war—are born by the majority citizen population. In fact, the security surrounding these "deal makers," as Nada referred to them, exists in stark juxtaposition to the precariousness of lives lived by most Lebanese, who must contend with the outcomes of political and economic crises fashioned by the power holders. In this sense, the insecurity wrought by formations of security during the 2004–2006 period represented broader, and enduring, experiences of and sentiments about feeling unprotected by an unstable and corrupt political system in a region rife with conflict.

THE INEQUALITY OF SECURITY

The havoc of insecurity and the differentiated experiences of mobility produced by processes of securitization during this time in Beirut recalled life during the protracted civil and regional war in ways both semiotic and material. It was not just the spectacle and destruction of the car bomb that brought back the war but also the fact that residents' mobility practices were once again regulated by armed forces, barriers, and closures, that territorial enclaves took shape, and that party loyalists-cum-foot soldiers working for the power holders of the political sectarian structure set up encampments throughout the city. And thus, while the order proffered by security was for the few, it created a kind of disorder for the many.

The establishment of security enclaves surrounding the residences of these individuals reconfigured the urban landscape in ways that entrenched its uneven geographies of power. As a result of these processes of enclavization, residents experienced forms of insecurity as they found themselves caught up in shifting and multilayered forms of security. After all, the mobility of the targets of security—important persons who move from place to place—meant that security formations too were mobile and thereby territorialized not only specific parts of the city but the city as a whole. Residents in the city, getting around by car and on foot, were enmeshed in disciplinary arrangements produced by material and social mechanisms of security. These mechanisms of security, operated by private nonstate and state actors working in concert, created an exclusionary urban landscape and expanded the privatization of the public realm.

5 · THE CHAOS OF DRIVING

In April 2006, I was sitting in the back seat of a service taxi when a man who appeared to be Filipino told the driver he needed to go to Sassine Square, a commercial area in Ashrafieh. The driver signaled for him to get into the car and he took the front seat. Right after we started moving, the driver swerved sharply and quickly to avoid hitting a car that had stopped suddenly. Speaking in very basic Lebanese Arabic, the front passenger made a comment about how crazy driving in Beirut is. The driver agreed, saying in reply that yes, there is chaos but there are no accidents (*fi fowda bas ma fi accidents*).[1] The passenger didn't agree: "Of course there are accidents; people get hurt all the time" (*akeed fi accidents, fi mowt, fi kasr*). The driver asked the passenger where he was from and what he did for a living. He was from the Philippines and worked as a cook in a restaurant. But these inquiries only led the driver down the same rhetorical path: that the chaos on the streets of Beirut is managed, that it is an orderly chaos, and, for this reason, there are few accidents:

DRIVER: Are there a lot of car accidents in the Philippines?
PASSENGER: Yes, more than here.
DRIVER: Exactly, there are more accidents there because the drivers are not used to sudden things happening on the road. But the drivers here are.

The driver is looking back at me through his rearview mirror, seeking support, agreement from me about the way things are in Lebanon.

KM (borrowing a phrase I just heard someone use): Well, it's true that the drivers here are clever.

During this conversation between the taxi driver and the front passenger, I sat in the back of the taxi silently agreeing with the passenger and refuting the driver's claims. I *had* seen many accidents in Beirut, and members of NGO groups with whom I had met in the course of my research confirmed that traffic fatalities and injuries were significant throughout Lebanon.[2] But I had also become familiar with this particular narrative about driving in Beirut, the one voiced by the service taxi driver, of there being a kind of orderly chaos on the roads that the city's skillful drivers knew how to successfully navigate. Moreover, no matter the ensuing analysis, the notion of "chaos" almost always emerged in discussions about being mobile in Beirut. After telling one service driver that my research was about "traffic," he responded, and offered his approval, by simply saying, "Oh, you mean 'the chaos.' That's a good topic."

Throughout this chapter, I make use of the word *chaos*, the standard translation of the Arabic word *fowda*, which Lebanese used to describe traffic in Beirut. However, I would like to make clear my disengagement from the notion that chaos is an objective and essential condition that can be empirically assessed and identified. Drivers and nondrivers alike spoke of this chaos. In our conversations, residents from different parts of the city, those from different socioeconomic backgrounds, men and women, young and old, all emphasized the chaotic nature of getting around Beirut. The examples they cited, the evidence of *fowda* they wanted to call my attention to, were not unique to Lebanon: they were examples of the frenetic and anything-goes nature of traffic, which resembled that in many cities around the world. But as I listened further, I realized that the discourse about chaos itself, rather than its practice, was especially significant. Amid these narratives of the chaos of driving in Beirut were understandings of critical issues in Lebanese social and political life, past and present.

SHARING THE ROAD: THE SERVICE TAXI

Although the streets are crowded with various types of vehicles, the service taxi dominates the scene. They are the most widely used form of public transit in Beirut. To flag one, passengers stand in the street, usually along a busy thoroughfare, and lean toward the window of an oncoming service taxi and tell the driver a destination.[3] In a city where numbered street addresses are not used and streets are commonly referred to by a variety of names, the destination is a landmark, neighborhood, or intersection. Usually, the driver will issue a nonverbal response on hearing a passenger's intended destination: by tilting his head and chin upward, he gestures no, or, issuing another form of silent rejection, he simply drives off.[4] Alternatively, by moving his head in the direction of the front passenger seat or toward the rear passenger area, the service driver issues his acceptance of a passenger. A service driver will continue to seek passengers as the taxi moves along, often until the vehicle is at a capacity he determines himself. For example, the driver may not pick up anyone else or the taxi may end up full, with three passengers in the back and one in the front.

Service drivers are always already engaged in the arithmetic of fare versus fuel costs that shapes their decisions about which passengers to accept and which to refuse. With no running meter—they are paid the same fare even if stuck in traffic—they anticipate possible traffic congestion in determining whether to accept a passenger. Sometimes, to save fuel, they turn off their engines while idling in a traffic jam. I remember, on several occasions, sitting in old Mercedes taxis that had trouble restarting after their drivers shut them off while standing still at busy intersections. Traffic would begin to flow after a time, and as the taxi struggled to get moving again, drivers behind us would beep and yell furiously.

In the first phase of my research, 2004–2006, the majority of service taxis were aging Mercedes from the 1970s and 1980s with no air conditioning; drivers said these cars were more durable and easier to repair than others. However, by 2010, many of these Mercedes had been replaced with cooled, more fuel-efficient Japanese and Korean cars. Indeed, gas costs ate up a significant portion of drivers' earnings, as I learned through informal conversations with drivers as a passenger in their taxis, and many told me that they made less than US$15 a day. Figuring just below the average per capita income level,[5] these wages created a strained existence: average family-sized apartments rented for between US$300 and $500 a month. Apartments in desirable buildings and in central or seaside locations rented for considerably more.

While living in Beirut, I saw only male service drivers, the majority middle-aged or older. There are those with and without an education, those who speak Arabic, English, and French—the three commonly used languages in Lebanon—and those who speak only Arabic. Moving around the city along a route they have fixed themselves, the drivers normally put in long, ten- or twelve-hour shifts. Beirut's streets swell with these taxis, and as the drivers head down streets beeping at pedestrians to signal that they are interested in picking up another fare, they produce one of the signature sounds of the city: a short, but significant toot. At first distracting, the tooting of the service taxis fades into the background of the public scene once one becomes accustomed to it. It is a poignant acoustic, an itinerant tale of trying to make ends meet.

I traveled by service taxi almost daily. Every so often, I took the public buses if I had the time to spare or wanted to save a bit of money. On occasion, I was a passenger in a private taxi or in a car owned by a friend, research participant, or the driving school where I went regularly to conduct fieldwork. For foreigners, using service taxis serves as an initiation into the geography of the city as well as into the specific mores that govern this mode of public transportation. For example, choosing where to hail a service taxi is crucial, particularly during times of the day when congestion is worse. Situating yourself on a street whose one-way traffic flows in the direction of your destination often gives you better luck in finding a consenting driver. Likewise, figuring out how to name a destination for the driver that is near enough to where you need to go but that is situated along an uncongested route so that the driver will not use up a lot of time and gas getting you there will also improve your chances of getting a ride with the very first taxi you hail. This kind of know-how about the city's layout and its most crowded corridors is an asset in using this transit system. If you lean down and tell the driver a destination that conjures images of bumper-to-traffic with no escape routes, it is not uncommon to find yourself rejected by one service driver after another, waiting in the street until you find a driver who will consent to take you where you need to go.

Like the roadside hand signals Czeglédy (2004) describes being used in communications between passengers and drivers of kombi shared taxis in Johannesburg, these actions represent localized strategies for navigating the service transit system. Adopting these strategies, and speaking proficient Lebanese Arabic, thus enables foreigners to demonstrate a kind of "indigenous knowledge" and to represent themselves, whether accurate or not, as residents of the city and not merely tourists. As elsewhere in the world, being identified as a resident, rather than a short-term visitor, by taxi drivers and merchants of all kinds also serves as a talisman against being overcharged.

MAKING SENSE OF CHAOS: CLASS

Service taxis play a significant role in the city's street life not only because of their sheer numbers but also because they are the site of intergroup interaction in Beirut par excellence.[6] Like other forms of public travel, service taxis are a social event in and of themselves, one whose mundane activities, as Paul Stoller wrote of the Songhay bush taxi in Niger, offer insights into the complexities of the social action.[7] As a locus of casual discussion about society, politics, and the nation among individuals from different backgrounds, the collective taxi serves as a kind of everyday civic forum as well as an example of Beirut's many scenes of public intimacy, which I wrote about in chapter 1. As such, encounters in service taxis are a regular conversation piece among residents of the city. And, in some conversations, they are also cited as being the chief source and orchestrator of the chaos on the streets.

Because services taxis do not pick up passengers at fixed points along the road,[8] and because they stop to engage verbally with passengers about their destinations when they do pick them up, other drivers cannot anticipate when, where, and for how long service taxis might stop. The service taxi's movement is characterized by inching along the right side of the road in the hunt for passengers and a sudden darting from the left to the right side of the road to do the same. At other times, the service taxi is stationary, in the middle of a lane of traffic, blocking the flow as its driver communicates with passengers standing along the side of the road.

A kind of anxiety is reflected in the service taxi's fitful movement, an anxiety born of economic stress. Ali Mohieddine, an official from the Taxi Drivers Syndicate, described to me in summer 2010 how taxi drivers were struggling and felt threatened because of the large number of illegal taxis on the roads. There were far more taxis than there should be, he said, because, "in 1994, the government passed a law allowing for the production of more red [taxi] license plates. So in 1994 there were 10,649 taxis [in Lebanon] and today there are more than 32,000. And this doesn't even include the copies—because the illegal taxis make fake license plates; they use the same number six times over!" Hearing these numbers prepared me for the estimates provided by Ilham Khabbaz, chief of Land Transport at the Ministry of Public Works and Transport, during our meeting that same summer. She estimated the number of service taxis on Lebanese roads to be about fifty thousand "though there should be only about thirty-three thousand," she added. Ms. Khabbaz spoke about how, in the future, decals with bar codes would be affixed onto the doors of service taxis and would eventually replace license plates as the legal requirement for public transit vehicles. The number of illegal taxis will be quickly reduced, she explained, because police will have devices to scan and read bar codes that are almost impossible to reproduce.

These issues of economic necessity, legality, and competition between taxi drivers were far from Nadia's mind, a university student in her twenties, when she described her commute to the university as a kind of obstacle course that culminated in—just before reaching the parking lot on the campus—an encounter with "the worst nightmare of all Lebanese drivers, a service driver: he stopped six times, in order to ask people on the street if they wanted to come in. He stopped without any sign, without even moving his hands up and down from the window to warn me that he was going to stop." In our conversation, another university student who drove her own car, Diala, spoke in a frustrated tone about how the service drivers "practically live on the road."

In this way, from the perspective of owners and drivers of private vehicles, the service taxi is understood to both incite and configure the so-called chaos of driving in Beirut. There is a class dimension to this understanding that has to do both with the very mode of driving as well as the social status of the drivers. For example, the taxi's movement, the perpetual going and stopping along the side of the ride to solicit passengers, is thought to create logjams that impede the flow of traffic. But it is also the lowly position of the service driver on the social and occupational hierarchy that makes his driving an easy target in explanations of the chaos, particularly from the perspective of those who own and drive their own vehicles. Service drivers have a high degree of public visibility—service taxis are recognizable by the taxi light atop the vehicle, the red license plate, or the way they move around the streets—but, at the same time, they constitute a disenfranchised social group. Visible and vulnerable, the driving of the service taxi is classed in ways that make them a kind of fall guy, easily blamed for the chaos.

Class also framed other forms of public transit. When I heard Noura, a young woman in her twenties who has her own car, say that she never takes service taxis, that her parents wouldn't allow it anyhow,[9] and that she absolutely never takes the bus because "we have this idea that the bus is for the foreign workers, for the Syrians," I was unsurprised. I had heard upper-income Lebanese offer this opinion of service taxis and buses several times before. Once, a wealthy owner of a publishing company even expressed bemusement about my research interest in public transportation as she concluded, "I don't really see why you are working on this topic; I mean, after all, there is no good public transportation here." I had my doubts that hers was an opinion based on firsthand experience using public transit.

Yet, Beirutis I knew who did not own their own cars used service taxis almost daily, depending on the sites of their workplaces and schedules, to get around the city. Some who did own cars chose to use service taxis or even public buses when convenient. Celine, a woman of similar age and background as Noura but whom I would describe as more politically engaged, had been given a car by her parents when she was a teenager. She described how she regularly drove to university but that sometimes she did take the public bus from her home in the northern suburbs. "I don't really mind it," she said; "it's a long ride, but it's cheap, and I can use the time to study." She described people she knew who would not ride the public buses because Syrian workers ride them. She dismissed these comments by tipping her head back so that her nose pointed upward to indicate that they were "snobby," as she put it in English.

For the owner and driver of a private car, the street—crowded with public transportation vehicles—can be seen as a field of impediments that constitute a kind of chaos. This type of chaos, the chaos that is instigated by stopping is, in this sense, linked with the realm of the less privileged and powerful. For the service driver, stopping involves solicitation, and for the passenger stopping involves having to wait on a public street for an affordable means of transportation. Both service driver and public transit passenger surrender some control over their movement through the city and directly interface with the open public environment and its myriad denizens.

Maya, however, asserted a kind of agency over her vehicular movement that seemed to contrast markedly with that of the service drivers and passengers. Maya, whose navigations of security I described in the previous chapter, usually asked me to meet her at her office in the afternoons. I would then go with her, and we would converse en route, while she ran various work-, child-, and household-related errands. To save time, she said, because parking in many parts of Beirut is so challenging, we often remained in the car—a black Mercedes SUV that she wanted me to know had been bought used—during these errands. Rather than park, Maya would call ahead to an office and someone would come out to the car and bring her what she had requested or needed to pick up, such as airline tickets or documents for work projects. She told me that she also picked up groceries in this same drive-by way by pulling up to a small grocery and leaning over toward the open passenger window to tell someone inside the store what she wanted. The items would be brought to her car, and she would pay for them there, while sitting in the driver's seat.

These drive-by errands invariably involved double parking, meaning she cut off a lane of traffic, but Maya was undaunted. While I had also seen other drivers of cars, including service taxi drivers, purchase coffee or vegetables from a street vendor in this same drive-by fashion, the way in which Maya called on others to bring things from within stores and offices to the car seemed different not only in degree but in kind. Her errand running demonstrated her efficiency in navigating the physical environment, via the insulation afforded by the private vehicle, to create certain kinds of opportunities for herself that were linked to the realms of both work and home. And what is more, even in her presumption of the possibility of store-to-car delivery service, Maya seemed to carry class privilege.

Maya's way of moving untrammeled through the city was suggestive of a different type of chaos from the one thought to be created by vehicles like the service taxi. Those double parking their privately owned vehicle in order to facilitate their consumer and professional activities or those driving too fast and without regard for others' safety might also be conceived of as chaotic. However, theirs is a chaos associated with privilege. Just as in other parts of the world, where the excessive driving speeds of young men who race through the streets are discursively constructed as a privilege of the elite, the young, male *shayfeen haloun* (show-offs) who speed around Beirut display their ability to move ahead, beyond, around.[10] They don't stop. Thus, in making sense of chaos, the movement of the service taxis and private vehicles is a site through which the geography of the city is classed.

OVERCOMING CHAOS: THE PROJECT OF DEVELOPMENT

Developmentalist notions that positioned Lebanon and the Lebanese as not yet "developed" or "modern" but certainly on the path toward these ends were also evoked in discussions of chaos. But it is perhaps at the driving school Traffic where I found ideas about "the chaos" being a signifier of the country's "development" status to be most conspicuous during my research.

Traffic, located on the ground floor of an office building in the downtown area, divided its curriculum into the French-termed *théorie* and *pratique*. The student-drivers I observed were teenagers, though there were also some adult learners at the school. The young people were a privileged group as evidenced by their English and French language fluency, their comments about traveling outside Lebanon, their stylish clothing, and their family's ability to afford private driving instruction.

In the théorie classroom, a loft space above the office area reached by ladder, I sat with the students watching French driver-education videos. The instructor stopped the video from time to time to go over finer points, the students asked questions, and there were moments in which the topic of what is specific about driving in Lebanon was raised. The théorie instructor would remind the students that once in pratique, they would find that French driving theory is often inapplicable to the reality of driving in Lebanon. Indeed, when I took part in pratique sessions as a rear passenger in cars driven by the students, the point most strongly emphasized by the teacher was to drive defensively at all times and to expect the unexpected. Though this instruction did not align with what the students had learned from the French driver-education videos, they appeared at ease with the idea that France and Lebanon were two very different places in which to drive.

Moreover, in the curricular culture of the driving school, France was positioned as a kind of role model and served a teleological function: French driving habits were situated in a distance that was ahead of Lebanon; they were habits that need be acquired in order to move forward and become more "modern" and "civilized." One driving student, born in Montreal but raised in Lebanon, spoke to me plainly about this trajectory: "You can see the state of a country in its driving—this reflects everything, the state of things." He gestured for me to look over at a car attempting to parallel park in a haphazard manner: "See, this is what I mean, this is a perfect example. You can see everything that's wrong or right about a country when you look at its driving. A country is not powerful because of how rich it is; its resources are its people, its organization. . . . This is Lebanon's problem!" In this young man's view, even if one could not exist outside of the chaos, one could at least—even should—recognize it as unproductive, if not a hindrance to national development.

Ziad and Mona Aql, directors of the Youth Association for Social Awareness (YASA), took another, less hyperbolic view of the relationship between the unruliness of driving in Lebanon and the nation's development. Founded by the Aqls in 1995 after the death of a friend in a drinking and driving accident, YASA was, at the time of my first phase of research, the only Lebanese organization working in the arena of driving and traffic safety. The organization provided educational and outreach services to citizens and the state alike. Through various activities including meetings and press conferences with government officials, public safety campaigns, and educational events staged at festivals and fairs, YASA urged Lebanese to slow down, to wear helmets and use child car seats, not to drink and drive, and, more generally, to abide by traffic laws and signage.

In conceiving of the chaos of driving in Lebanon as a grave public health issue that requires the combined effort of citizens and the state, Mona Aql framed the issue for me in developmentalist terms: "The thing is, as a developing country, like in all developing countries, traffic safety is not that big of an issue. It's not a priority. We have issues with enforcement, with accountability, with road engineering, issues of security. . . . We are getting there but so far injury prevention has not been a top priority. We need to lobby more, we need to make the government more accountable for this, but the accountability issue, that's a problem too in all developing countries."

The picture of the nation that emerges in Mona's perspective on chaos is one in the process of becoming. For Mona, the disorderliness of driving was a signifier of Lebanon's development status. This idea, that the chaos of driving was a behavior that would be overcome by progress, emphasized the everyday dimensions of the broader-scale processes that we normally think of as constituting national development.

This notion of development as a project related to everyday social practice departed from the hegemonic vision of development in the era following the civil and regional war (1975–1990). This was a vision for Lebanon's progress that focused on economic growth, which was to be achieved through the expansion of the banking and services sector and the physical construction of high-end commercial, residential, and office buildings. In conversations about the traffic situation, some people I spoke with voiced criticism about this model for Lebanon's postwar recovery. What was missing from this recovery plan, according to Dr. Hanna El-Jor, a public health consultant and YASA board member whom I spoke with in fall 2004, was an "investment in the people." The lawlessness one sees in driving, he said, is a reflection of the failures of the postwar reconstruction project.

During a service taxi ride in summer 2010, the driver echoed this sentiment. Most service drivers, upon hearing about my research about driving and traffic, offered their opinions about the subject readily. This driver was no exception as he described how "the newer generations learn from the older generations so the only way to change things, to change the system, to change the 'chaos,'" he joked, "is by taking away all the people that are here and replacing them with a new people!" His comments made me think, once again, about the interest people seemed to have in talking and sharing their perspectives about driving behavior. Part of this interest might have stemmed from the fact that, on the face of it, driving seemed to be a relatively uncontroversial topic that had little to do with the contentious sphere of Lebanese politics.

But driving also seemed to be an experience rife with emotion, one, as I describe in the following section, residents of the city drew on to comment on civic life.

THE CULTURE OF CHAOS: LEBANESE-NESS

Conversations about the chaos of driving were also linked with national culture and identity. Chaotic driving, in short, was often described to me as a demonstration or performance of Lebanese-ness that was recognizable among Lebanese from all walks of life. In *The Autostrad: A Mezé Culture—Lebanon and Auto-mobility* (2003), a research project and publication developed by the Faculty of the School of Architecture, Art, and Design at Notre Dame University in Louaize, a northern suburb of Beirut, the disorderliness of driving in Lebanon was even likened to another aspect of Lebanese culture: food. Getting around, the comparison went, was like Lebanese *mezé*, "a typical Lebanese meal wherein a large number of small hot and cold dishes are placed haphazardly all over the table, sometimes even on top of each other due to lack of space" (2003, 1). Moreover, in expressing their Lebanese-ness through driving, some respondents thought a sense of civic and national solidarity was effected. For instance, as Ali, a chauffeur and assistant for an upper-middle-class family, explained: "Despite everything, despite all of the divisions and the problems in this country, . . . the thing is, we [Lebanese] understand each other on the road [*nafham ba'ad a tareeq*]."

Ali's understanding of the chaos of getting around was not the only one that struck a collectivist tone. Others also described being mobile in Beirut in ways that were reminiscent of Anderson's idea about how the nation is conceived, and discursively constructed, as "deep, horizontal comradeship" (1983).[11] For instance, when I mentioned to Lara, a graduate student in her early twenties, that some people had told me that contemporary driving behavior stems from the era of the protracted civil and regional war, she disagreed. "No, no," she said, "it's the Lebanese way." I asked her what this meant. "We like to be clever [*shatreen*], driving our own way, finding a way to the front of the line. . . . We like this. . . . We are Lebanese." She added, "We do it for fun." She continued by comparing driving and standing in line. "When I went to France, I didn't stand in line. I would be at the end of the line, and I would see someone I know toward the front and go up and talk to my friend; and then I'd be at the front of the line. Sometimes I'd be with someone else and I'd grab my friend and say, 'Let's go to the front of the line.'" I told her how infuriating this behavior was for me, that I had encountered it at the bank, post office, pharmacy, and other places. Laughing, she replied, "I know, I know, but it's just how we are." But why find a way around, why do people do these things? I asked her. "We do it for fun; . . . we do it to show . . . that we are able to do it. To find a way around, a way through, it's the Lebanese way!" Another young woman, Dima, concurred with this analysis: "You know, Lebanese complain about the 'chaos' but they like it. They like being able to navigate around; they like the challenge, and they especially like not following rules!"

I also heard Lebanese driving behavior compared to that in other parts of the world in ways that echoed Lara's celebratory narrative about the Lebanese maverick sensibility being observable through driving practice. For example, in late 2005 while I was riding in a service taxi, an army soldier entered and took the front seat next to the driver. We stopped at a large intersection, where a policeman was managing the flow of traffic. While the traffic policeman screamed at the driver of a car that had gone through the intersection without being directed to go, the soldier amusingly remarked to the service driver, "You know, in America, they stop at the intersection even if there isn't a light!" The service driver laughed. This commentary and others like it seemed to make reference again to a developmentalist discourse about where Lebanon is positioned vis-à-vis Western countries but in a way that did not express feelings of inadequacy, but rather a kind of insider-ness and self-mocking pride.

Samah, a young man in his twenties who was four years old when his family emigrated to France during the long war, also spoke about how insider-ness/outsider-ness is expressed and cultivated through driving behavior. "When I come back to Lebanon, my friends are always saying that I'm *nizami* [orderly/law-abiding] because I follow the [traffic] rules." In another example, a social science professor at a university in Beirut, inspired by the many conversations we had had about my research, created a final exam question that asked her students to discuss driving in Beirut using concepts from the course materials. In their responses, many students made reference to cultural norms and values, but, more than anything, the students wrote about their sense of the sheer chaos on the streets and offered interpretations of its cultural meaning. One student, for instance, wrote about how driving behavior was a means of performing Lebaneseness: "Even if foreign people are not used to such a way of driving, they have to learn how to drive in Beirut. So, by driving between cars, not stopping at a red light, the person who is driving will be described as the 'real' Lebanese driver even if he/she is not as 'real' or from another country but he/she will look like 'real.'"

Sitting on a bus returning to Beirut from a summer concert in the mountains in 2005, I listened in on a conversation between what I surmised, as a result of their ease alternating between fluent Lebanese Arabic and American English, were two young Lebanese American men. Like the student's exam answer, they talked about how a certain set of skills was required for driving in Lebanon. "The thing about driving here," one guy told the other, "is that if you can drive in this country, you can drive anywhere." In a commentary for the *World Affairs Journal*, American journalist Michael Totten (2005) elaborated on this sentiment by describing driving in Lebanon as a game in which "your reaction time—and therefore your driving skill—grows exponentially after you've played this game for a while."

The theme of native or, for nonnatives, acquired, expertise and knowledge about how to transgress was a recurring one in the accounts about driving that I gathered during my fieldwork. As Lara described it, this was expertise in finding a way around, a way through, a way out. It was an expertise born from the cultivation of a particular set of skills of negotiation, in this case negotiation of Beirut's urban space, which suggested not only cleverness but also resilience in the face of challenge. These aspects of knowing how to drive in Lebanon went along with the ability to identify and to make the most of opportunities and with irreverence toward forms of state authority that seek to limit or constrain public behaviors and practices. Thus, together with social class and the developmental trajectory of the nation, I found that at stake in the talk about the chaos of driving in Beirut were matters of national culture and identity. In ways that evoked Anderson's (1983) idea of the nation as an "imagined community," in which dimensions of social solidarity, rather than inequality and exploitation, are emphasized, people of varied backgrounds described the competence and skill said to be required for driving in Beirut as expressions of Lebanese-ness.[12]

In hearing these comments about getting around Beirut, I began to think about how driving, conceived of as practice (in the Bourdieuian sense of an action with a history), could also be understood as *habitus*, a disposition, as Pierre Bourdieu wrote, that is "collectively orchestrated without being the product of the orchestrating action of a conductor" (1977, 72). In the narratives about clever and self-determining driving behavior being Lebanese, everyday mobility emerged as a medium through which a kind of national habitus could be observed.

THE CULTURE OF CHAOS: DISORGANIZATION AND MORAL AND CIVIC PROPRIETY

Lebanese draw on the notion of *fowda* not only in their talk about driving and traffic but also in their descriptions of other dimensions of civic, political, and everyday life. Ideas about the chaos of being mobile in Beirut are thus situated amid broader understandings of social disorganization.[13] For example, I heard activities as diverse as infrastructural improvements, mundane dealings with the bureaucracy, and queuing practices characterized as disorganized or chaotic. One respondent described how construction projects in Beirut are always ongoing and never quite finished because the different utility and service companies do not coordinate their installation projects with one another: "The telephone company digs up holes in the street for a month to install cables and then the holes are closed up; . . . the next month the electric company comes back and digs up the street to install something. So the street is always torn up!"

Religious scripture offered another lens on the moral and civic sensibilities surrounding driving and public behavior more generally. Grand Ayatollah Mohammad Hussein Fadlallah, the now deceased leading Lebanese Shi'i Muslim cleric and spiritual leader of Hizbullah, gave the subject his attention. In 1995, he issued a *fatwa*—or formal legal opinion or decree—that called for adherence to the traffic law and the preservation of public space for safe and unrestricted civic usage. "Compliance with the traffic system," he wrote, "will put an end to a chaotic driving which leads to the disorderliness of not only public space, but all aspects of people's lives."[14] In the same *fatwa*, he also discussed the necessity of taking care of public space so as to ensure the safety of residents by advising, in particular, that owners of stores refrain from taking up sidewalk space outside their shops so that residents are not forced to walk in the street to get around the display of wares. He wrote that it was "not permitted for one person to behave in the streets in a way that creates problems for the movement of others, be it [by] parking cars so that streets are turned into garages or by creating obstructions that are not due to cases of emergency." More recently, in 2007, Fadlallah issued a decree prohibiting the obtaining of a driver's license by illegal means and, in the context of driving, insisted that Lebanese respect their health and safety and that of others.[15] Just prior to his death in July 2010, he met with the traffic-safety organization YASA and "affirmed the legal Islamic position that requires respect for public order and recalled his earlier *fatwa* stipulating the necessity of respecting the traffic laws."[16]

These decrees conceive of driving as an act of moral citizenship. And, as part of the civic realm, the ways in which residents move through urban space—and prevent others from moving through it—are thought to belong, in Fadlallah's words, more to the realm of obligation and responsibility than to the arena of rights and entitlements. A sense of constraint that comes from within, distinct from the external constraints exerted by the police and the state, emerges here: the moral citizen should herself strive for decorous driving, rather than expect others to impose it. The Islamic conception of *fitna*, meaning upheaval, disturbance, strife, is brought into play here. The individual must engage in a moral, internal struggle against *fitna* that is connected to a broader, external struggle against a more publicly related *fowda*.

Along with these religious framings of the disorderliness of driving and other kinds of public behavior, many residents of the city offered understandings of the chaos that drew on a secular framework of civic responsibility and ethics. Samar, a project manager at a nonprofit organization focused on citizenship education for children and youth, spoke about how upsetting she found the driving situation in Lebanon. While some seemed to celebrate the anything-goes style of driving, she viewed unruly driving behavior in a definitively negative light: "It really upsets me, and I'm sorry to say this because I'm Lebanese, but it really shows me when I'm driving how the Lebanese are, how they are opportunistic; it shows their ethics." During our very first meeting, when I introduced my research topic to Maya, the working mother who ran errands by driving by, she offered a similar commentary when summarizing what she saw as the core issue undergirding the chaos: "We are enemies on the road. It's a competition. Everyone wants to be first, . . . and even professors and people I know who supposedly have different ethics about citizenship, about civics, they do the same thing on the roads. 'I have to be first'—we all do it."

THE CHAOS OF CORRUPTION

During my research, I found that in addition to matters of civic responsibility and ethics, corruption was also a part of everyday parlance linking the chaos of driving with broader structural issues of social disorganization.

In July 2010, a Lebanese television news program, *Al-Fassad* ("corruption" in Arabic), featured as its main guest Ziad Aql, director and co-founder of YASA. Throughout his interview with the program's combustible host and in responding to questions from callers, Aql drew connections between chaotic driving and forms of government corruption. "In a system where we have twenty cars driving around with the same license plate number and people getting driver's licenses without taking the test," he argued, "it is not surprising that we have 'chaos' on the streets. The numbers of accidents, the problems with traffic, these are problems of corruption."

While the television program *Al-Fassad* provides a platform for citizens to air grievances about corruption in and beyond the realm of traffic, the Lebanese Transparency Association, a nonprofit organization, focuses on advocating for reform in governance and citizen behavior. In a meeting I had in 2010 with Badri Meouchi, the association's executive director, he detailed their three key steps: raising awareness, doing research about how to improve the system, and making recommendations to the decision-makers about the changes that should be made. When I used the phrase *culture of corruption*, Meouchi bristled: "When you say culture it sounds like something that can't be changed, something that comes from families and traditions.... It's better to think about it as a question of habits, of practices. Let's take the parking meters, having to park inside white lines and then get out and go and pay at a machine! I mean, come on, ... this is something you couldn't ever imagine Lebanese doing! But now they are doing it. And they like doing it!"

In this sense, to take Meouchi's point a bit further, the chaos in Beirut could be contested through awareness, education, and the incorporation of new habits into everyday life. Curiously, in fact, Meouchi's comments about Beirut residents' reception of the parking meters, particularly given that any form of parking serves as a constraint on unfettered automobility, provide a counterweight to the discussions of Lebanese-ness that argue for the cultural authenticity of a mobility that breaks all the rules.

The pay-and-display electronic parking meters Meouchi spoke of were a recent addition to Beirut's streetscape. The meters, along with the construction of pedestrian bridges and the installation of traffic lights, were part of the Urban Transportation Development Project (UTDP), an infrastructural effort funded jointly by the World Bank and the Lebanese Council for Development and Reconstruction (CDR) to improve traffic conditions in and around Beirut. I met with the manager of the UTDP in Elie Helou in 2005 and again in 2010 at what marked, respectively, the early and final phases of the project. In 2010, after a conversation during which he spoke with pride when detailing the infrastructural improvements that had been completed and with frustration about the state's inability to successfully manage and administer the traffic system that makes use of these improvements, Helou walked me from his office to the front of the building housing CDR in downtown Beirut. Helou, a garrulous presence, waxed contemplative when I asked him about the lessons learned through his involvement with the project. Echoing Meouchi's sentiments about unforeseen citizen behavior, Helou paused a bit before saying, "You know, I really thought," then, correcting himself, he continued, "No, *we all* thought that the parking machines were going to be the weakest part of this project, . . . that there was no way that people would use them, that people would hate them and complain about them all the time. But it turns out they are the strongest [part of the project], the most successful part. People really like the parking machines."

Meouchi's and Helou's comments about residents' use of and satisfaction with the parking meters gave support to the idea that the chaos of driving in Beirut was, at least in part, the outcome of a disorderly government that took little, or ineffective, action in ameliorating the traffic situation. Moreover, the order imposed by the parking meters was one welcomed by citizens, perhaps, I want to suggest, because they constituted a transparent, efficient, and predictable system. The locus of the system, in the case of the pay-and-display parking meter, was a technology that seemed to prove both gratifying and empowering for its users. Although other traffic-related infrastructural improvements, such as new roads, pedestrian bridges, and even traffic lights, gave the promise of efficiency, stability, and predictability, their success relied significantly on the efforts of the state actors to manage how these roads, bridges, and lights would be used by drivers and walkers. The parking meters, though also a technology of government that was part of the engineering of public space, afforded citizens a more direct relationship with civic order and regulation, one that was achieved through interface with technology rather than interaction with the human face of government. The parking meters, in a sense, created the possibility of a different politics of mobility in Beirut. In this way, sentiments about governance were manifest too in discussions of a chaotic mobility in Beirut.

Maha, a woman in her fifties who raised two now-grown daughters on her own, in summing up her feelings about how the state should deal with unruly driving, invoked the roles of parent and child: "The government is responsible first, not the citizens. . . . It's like parents and children, the parents have to show the children the right way." To extend Maha's parent-child analogy a bit further, in the context of everyday mobility, an ill-functioning relationship between citizens and state was fostered by feelings, on the part of citizens, of not being taken care of by the state.

MOBILE AND DIFFERENTIATED CITIZENS

In *Flesh and Stone* (1994), Richard Sennett traces the historical development of the Western civic body. The movement of this body, set in the urban context, is a desensitized one, he argues; it is a body that experiences space "as a means to the end of pure motion" (1994, 17). In the Western context, he writes, spatial forms and practices that limit contact between bodies—and citizens—are both sought out and achieved through mechanisms of planning as well as experiences of spatial movement. This body, Sennett conceives, is passive rather than active, and the urban, Western citizen who inhabits it "wants simply to go through urban space, not to be aroused by it" (1994, 18).

In Beirut, I observed a different kind of civic body. Entangled in the chaos of getting around, the civic body in this setting was exceedingly active and, what is more, many of the Beirutis I spoke with during my fieldwork understood their movement through the city to be a meaningful aspect of their lives as citizens of the urban and national space. These understandings, I suggest, demonstrate that driving (or riding) is an everyday experience through which different Lebanese express and recognize shared interests and sentiments about the development of the nation, civic culture, the political and governmental structure, and the national self. And yet, while driving or riding was a practice performed by all, vehicular mobility was also an arena for social division and differentiation. While I heard consensus about the presence of a kind of chaos, for example, the movement of taxis and other public transit vehicles was often cited by those in positions of relative privilege to be both a producer and a product of the chaos. In this way, driving, as a specific form of spatial movement, had to do with one's right to take up public space and was an entitlement possessed unequally by Beirut's residents.

In their talk about the chaos that characterized getting around Beirut, many residents described a national identity that involved the possession and demonstration of a maverick sensibility in the face of state ineffectuality, neglect, and corruption. The city, in this delineation of chaos, becomes a stage for the collective rhythm of different social groups, a site of alliance.[17] However, narratives of getting around Beirut also revealed a city mapped by hierarchies of class and power. The mobility of the service taxi driver, for instance, was talked about by some as being the source of the chaos, and his lowly status and occupational maneuvers were thought to be a root cause of the "problem." Those who own or drive private vehicles can get from point A to point B more quickly than those who take other forms of transport. But along the narrow, and often one-way, lanes of Beirut, the movement of service taxis and buses often impedes the movement of private vehicles. Thus we find, in the discourse about chaos, a key contradiction. On the one hand, talk of getting around brings together different kinds of Lebanese under a particular political imaginary: "clever" and "competent" citizens coping with a "developing" nation and incompetent state. On the other hand, distinctions among citizens, and their stratification, are engendered by this same discourse. Like navigation through the "secured" city, which I discuss in chapter 4, an urban citizenship differentiated by mobility and spatial access is also produced through talk of a chaotic getting around. In this way, being mobile in Beirut is a significant civic practice, one in which order and disorder commingle and various forms of boundaries, both social and territorial, are fashioned.

6 · "THERE IS NO STATE"

On a Sunday afternoon in June 2010, along the main road in Mar Mikhael, an eastern Beirut neighborhood, a traffic patrol (*dowriyya*) was set up by the Internal Security Forces (ISF) to catch helmetless motorcyclists.[1] This attention to and enforcement of the normally flouted helmet law was part of Interior Minister Ziad Baroud's public safety campaign. Baroud's term began in 2008 during a tense standoff between the March 14th and March 8th political camps that had turned violent and brought urban combat to the streets of Beirut and the nation to the brink of civil war.

Within this broader context of unsafety, two ISF policemen, acting on the ground to impose Baroud's policy priorities for road safety, laid a trap. One policeman, dressed in ISF's gray camouflage, hid in the entryway of a building positioned along a curve in the road where he would go unnoticed by oncoming drivers. About twenty meters down the road, within the line of sight of the soldier, a policeman, wearing the traffic-police division's blue uniform, stood in the street beside his motorcycle. On my way somewhere, I stopped to see what was going on. As a couple on a motorbike sped past the hidden policeman, he turned toward and signaled to the other policeman by pantomiming whistling with his fingers in his mouth. The policeman pulled the couple over and began to issue a ticket. Soon after the couple drove off, two young men on a motorbike drove toward the policemen, but, at a split in the road, an SUV slowed and its driver spotted the patrol. He beeped his horn to give notice of what lay ahead, and the young men turned around and sped off.

What struck me about the patrol I observed that afternoon is how it recalled the sentiments held by traffic police who spoke to me about Beirut's citizen-drivers' irreverence toward their efforts as agents of the state to regulate everyday mobility. The police had arranged a game of cat and mouse, but who was the cat and who was the mouse? Relying on the tactic of surprise attack, the policemen put their bodies into their work as one stood exposed in the street and the other used perhaps the most rudimentary of equipment, his hands, to set things in motion. After this orchestration, civilians working together disregarded and easily evaded the trap police had laid for them. When the guys on the motorbike effortlessly fled the scene, I even felt a little sorry for the agents of the state because they were the ones made to appear vulnerable out in the public view.

Although this scene of citizens banding together to evade police regulation can be observed in countless sites around the world, in the context of Beirut it was shaped by particular dynamics of policy and law enforcement that constituted the leadership priorities of Interior Minister Baroud. From the start of his tenure, Baroud vowed to get tough not only on violators of traffic laws,[2] but on crime at all scales. As an attorney and low-profile figure appointed by President Michel Suleiman, a conciliator between the March 14th and March 8th camps, Baroud took a ministerial approach steeped in what anthropologists Jean Comaroff and John Comaroff have described as a globalized culture of legality whereby crackdowns on criminality are asserted as the means and ends of governance. While the Comaroffs (2007, 136) describe how this culture of legality has appeared in response to emergent forms of profiteering by illicit and licit businesses in an era of neoliberal restructuring, as zones of deregulation become spaces of opportunity, inventiveness, and violence, Baroud was focused on reining in everyday forms of lawlessness along with corruption among power holders who understood themselves to be above the law and who carried on business as usual amid the machinations of sectarian-based politics.

In staging his crackdown, Baroud took a hands-on rather than solely discursive approach. Stuck in traffic in downtown Beirut in December 2009, he famously exited his car and began to try to improve the situation by removing barriers that were blocking roads. Afterward, he made public his letter to the ISF condemning their "closing public roads in the absence of administrative approval" from the proper authorities (Rizk 2009). This incident was only the first in which Baroud presented himself as a kind of Everyman by standing in the streets directing traffic, and it presaged future, and irreconcilable, differences between himself and the head of the ISF, General Ashraf Rifi. In its encouragement of adherence to the law, his public stance was avowedly anticorruption and was aimed at political elites and regular citizens alike. Through word and deed, Baroud aligned himself with the lowest ranks of the ISF—the men directing traffic at intersections— and he included officials among his targets of civil reform. Like ordinary city residents, Baroud thought the chaos of traffic in Beirut was not only revealing but also constitutive of a kind of disorder at the broader scales of society and government. Although residents may not have exactly echoed Baroud's discourse of legality as the antidote to this disorder, I found that in their comments about "the state," they shared his raw feelings about the failings of structures of governance.

In this chapter, I show traffic encounters in Beirut to be a key site for the production of the state by focusing on how, through the policing of vehicular movement, members of the traffic police perform a kind of embodied work that is part of the everyday formation of the state and, in this respect, how civilians' experiences with and ideas about traffic enforcement draw out understandings of the state as corrupt and ineffective. From this field of urban mobility, I consider how the state is constituted by the perspective of both citizens and street-level bureaucrats, who, despite, but also as a result of, their discord, share common ground in their sense of insecurity and disappointment with the state.[3]

BEYOND "THE WEAK STATE"

The persistence of clashes between political groups in Lebanon and the growing incorporation of Hizbullah into the government are the kinds of examples scholars of international politics, security studies, and policy makers alike use to characterize Lebanon as a "weak," "failed," or "fragile" state that has little military might and is susceptible to being overrun by internal, nonstate "terrorist" organizations with transnational links.[4] Many of these theories of the weak state draw on a Weberian definition of the state as a single entity that can claim a monopoly on the legitimate use of violence over a given territory.[5]

The understanding of the Lebanese state I present here seeks to move beyond interpretations of the state as weak or failed. Instead of pursuing a line of thinking about the state as a unitary "thing" that is categorically weak or strong, I ask, What might traffic encounters of and between police and citizens reveal about the state? By taking seriously the sense of powerlessness and insecurity both the traffic police and citizens attribute to the flawed functioning of the state "that is" as well as their expectations for the state that "could be," I hope to offer an alternative to the weak-state discourse by showing how state power is contingent upon various kinds of social relations that are worked out in—and productive of—civic life in everyday ways.

The traffic encounter is a localized and unspectacular site through which the idea of the state is mobilized. Moving beyond an exploration of how state projects of legibility are enacted, I show through my ethnography how citizens actually meet these projects head-on in their interactions with traffic police in Beirut. In the context of Lebanon's disordered political field and unfulfilled hopes for a better-functioning state, I see these interactions less as the foot-dragging measures of everyday resistance that Scott described (1985) than as outcomes of disaffection with the state. Moreover, I illustrate how not only citizen-drivers but the traffic police as well convey these feelings of disaffection. Although we often think about state power in ways that consider subjects and agents of the state as adversaries, I offer a different picture of the relations between these two groups, one in which they in fact share sentiments about the state and about the state of things.

I begin with an inquiry into how residents' understandings of the state as ill-functioning, demonstrated by the oft-used expression *ma fi dowla* (there is no state), emerge from the "chaotic" traffic encounter. I then elaborate this account of the state by shifting from the perspective of citizens to that of agents of the state by looking closely at the workaday experiences of the traffic police and their endeavor to show the force of the state in spite of the challenges to its authority they endure. Finally, I delve further into the complexities that surround police work by exploring how the wielding of *wasta* (connections) makes matters of class and status central in state projects of law application and enforcement. I conclude by considering what these perspectives of the traffic police and citizens tell us about their conceptions of the state. Curiously, feelings of disaffection and being unprotected emerge as key themes of civic concern for both police and civilians as they reveal expectations for an as yet unrealized, different kind of state.

THE ILL-FUNCTIONING STATE

As I discussed in chapter 5, when I asked people about the source of the so-called chaotic traffic conditions, the long civil and regional war was often referred to. "It's because of the war," I heard, and people would go on to say that reckless driving today is connected with the collapse of law and order during the civil war years. Even people too young to remember use the temporal frame of "before the civil war" to compare contemporary life with the way things were. Residents conveyed their sentiments about the ineffectuality or breakdown of the state through these references to the civil war, and, while in service taxis, I often heard passengers and the taxi driver criticize the police specifically, and the state more broadly, for allowing the chaos on the streets to go unchecked.

When a garbage truck emptying dumpsters blocked traffic on a narrow thoroughfare during rush hour, for example, the occupants of my taxi not only blamed the trash collection company's disregard for drivers but agreed that the police too were at fault as they were never around when problems like these arose. And while a global discourse emerged about the buzzing noise made by fans blowing horns called *vuvuzelas* during the 2010 World Cup in Johannesburg, many Beirutis I spoke with that summer also complained about noise generated by the late-night-to-very-early-morning antics of young Lebanese caravanning around the city bearing the flags of their favorite teams, incessantly beeping car horns, and setting off fireworks after the matches had concluded. The police, they said, do nothing to control this behavior, even though it is not only disturbing but unsafe. Drivers I accompanied, in buses, taxis, and private vehicles, would issue vitriolic complaints not only about the drivers who double- (sometimes triple-) parked along corridors, blocking the flow of traffic, but also about the state that allowed this kind of public behavior to go unchecked. Although Lebanese commonly greet policemen and soldiers with the respectful and patriotic appellation "[the] nation" (*watan*), in their talk about traffic and the state, they indicated disapproval, even disavowal, of the institutions charged with security and safety for the ordinary citizen.

Once, sitting in the backseat of a taxi, I listened to the driver, a man who looked to be in his fifties, and his front passenger, a woman of similar age, discuss the bottleneck that had brought the car's movement to a stop. After their conversation trailed off with mutual agreement about the fact that "there is no state," I asked, "When *was* there a state?" The question seemed to catch them both off-guard, as it interrupted their focus both on the immediate present—sitting in traffic in a hot, crowded taxi—and the broader temporal context of contemporary life. "Before," they went on to say after I pressed further, there had been a special police brigade that controlled traffic and no one dared disregard their authority. They didn't remember exactly when this was, that such a brigade existed, but it was during a time "before the civil war" they said.[6]

Curiously though, as I show in the following section, the traffic police had emphasized to me their struggle to effect order through, for example, getting residents to comply with traffic rules. In particular, their attempt to create order out of the chaos was challenged by those who aimed to use their status, or their connections to someone else's status, to disrupt, essentially, the effectiveness of the state. How do we reconcile these two seemingly contradictory understandings of vehicular mobility, that of a beleaguered police unit recounting their difficulties in enforcing the law on the one hand and civilians' descriptions of the state as ineffectual on the other? One way, I suggest, is to look closely at these understandings for the expression of shared sentiments about the state. Both groups, the police and the residents, articulate feelings of insecurity about the state and about the state of public and civic life.

This notion of sentiments about the state being shared by both civilians and police framed one of the more surprising moments of my research. Sitting in the public relations office of the ISF, where I had come to apply for official permission to conduct interviews with members of the traffic police, I talked with the officer—his uniform seemed to indicate that he was of officer rank—responsible for handling my request about my research on traffic in Beirut. He began commenting about how unruly the situation was, and, somewhere in the middle of his remarks, trying to explain to me what the root of the unruliness was, he observed summarily: "Well, there is no state." While initially caught up in the crisp irony of his statement, I later considered the way in which he used this phrase, as did residents getting around the city, to express how traffic problems demonstrated a system that was not working as it should.[7]

The question begged is, Why do Beirutis regularly use the expression "there is no state" and not "there is no government" to decry the chaos on the roads? The answer, at least in part, lies in the broader problematic about the state that is brought into play by the ways in which people make sense of the traffic encounter. The hazards of driving in Beirut—cars traveling at breakneck speeds, drunk drivers, unexpected road obstructions, poor lighting, double-parkers blocking the flow of cars, drivers heading the wrong way down one-way streets—are concrete illustrations of the kinds of everyday and routine dangers that people cope with while being exposed to the risks posed by the outbreak of political violence and war.

Moreover, when Beirutis express feelings of being insecure and unprotected through their comments about the state's ineffectuality with regard to traffic, they tie citizen safety and security—certainly in a corporeal but also in a more existential sense—to the apparatus of the state, as opposed to the workings of government. In short, it may be important for us to take note of the ways in which the categories of state and government are not interchangeable at a conceptual level because, quite simply, people themselves rely on and find meaning in this distinction.

As Kivland (2012) observes about Haitians' notions of statelessness, Beirutis' claims that "there is no state" are expressions of a desire for something different. Traffic is an immediate and concrete realm of public life through which people express unfulfilled expectations for a state that might serve as a protector, and a provider, for ordinary citizens. This is a realm of life that is always already tied up with myriad other issues including the instability of and corruption in the political system, economic insecurities, poor infrastructure, and inadequate public services. In this way, people's use of *ma fi dowla* registers concerns that go beyond the specificities of traffic to convey feelings of dissatisfaction with an ill-functioning state that leaves them to fend for themselves.

TRAFFIC TRAINING

Hierarchically arranged around rank, supervisory role, and work duties, employment with the traffic-police division of the ISF, staffed primarily by men ranging in age from their twenties to late forties, is a stable middle-class job made attractive by the social security, benefits, and relative prestige it provides. Within the broader structure of the ISF, which directs all domestic security matters from staffing at foreign embassies and the issuing of visas to conducting intelligence operations, the traffic police play what might be understood as a modest role within a highly significant government agency. In a postwar era rife with political crisis, violence, and assassination and in a city inhabited by both visible and covert paramilitary groups that include the privately hired security forces of VIPs, armed civilians-cum-militias, and foreign mercenaries, the ISF manifests the Lebanese state in the day-to-day securing of public space.

The ISF is one of several departments of internal security under the helm of the Interior Ministry that constitute competitive fiefdoms identified with particular sectarian groups; the Bureau of Intelligence is thought to be Sunni-controlled, for example, and the General Security branch, Shi'i-run (Belloncle 2006). From 2005 to 2013 the ISF was led by General Rifi, a man who hails from the predominantly Sunni city of Tripoli, sharing a hometown with then–prime minister Najib Mikati (prime minister during particularly turbulent political times from April to July 2005 and from June 2011 to March 2013). The ISF's gendarmerie might act as the traffic police—pulling over a driver who has committed a traffic violation, for instance—but members of the traffic police do not undertake the domestic-security activities of the ISF's gendarmerie. In this way, the traffic police are situated squarely within the state's realm of policing through the maintaining of public order and the display of expertise, and, yet, they remain on the margins of the security apparatus as a whole.

Once, while meeting with a staff member from YASA, the traffic-safety organization, I was surprised to learn that YASA also provides training to the traffic police. I wondered, Why do the traffic police need to be educated on the importance of traffic safety? This was a question I sorted out a preliminary response to after attending a YASA-led "class" for traffic-police trainees.[8] In a tired-looking army administration building in Baabda, just outside Beirut, more than one hundred traffic-police trainees dressed in gray and white fatigues filed into a large multipurpose room and stood in front of wooden chairs with desk arms attached. Facing them on a stage, a soldier signaled that they should take their seats. He introduced the YASA representative and, over the next hour and a half, the young men listened to a driver-education lecture that began with the following message: "I'm not here to flatter you. I'm speaking the truth when I say that you are in the privileged position of being responsible for the safety of society."

As the lecture ran past the half-hour mark, the men grew restless.[9] Semi-muffled conversations could be heard from where I was sitting at the front of the lecture hall. Some gave up feigning interest altogether and laid their heads on the desks. They livened up when the question-and-answer period began. With the exception of the first question—"What's a safe driving distance between cars?"—the trainees all gave examples from their own driving-accident experiences, seeking the lecturer's assessment of who was at fault. The soldiers were invigorated, talking over one another to weigh in. As I listened, it seemed that what animated the young men most was a sharing of testimonials that ended in a determination of blame. Sitting in the front, I could see the lecturer from YASA, in his responses to their queries, becoming frustrated as he struggled make this a teaching moment whereby their individual accounts might be turned into a civic lesson about the responsibilities of all drivers. I considered how YASA's role, in providing training to the traffic police, was to arm them with a sense of the significance of their enterprise, as evidenced by the lecturer's opening remarks about their being responsible for the safety of society. By elevating their position in the state security apparatus, the lecturer's message seemed to be one not only of empowerment but also of accountability; it was a message about duty and honor that the trainees might carry with them to the field of challenging vehicular, as well as human, relations when managing traffic in Beirut.

SHOWING THE FORCE OF THE STATE

YASA was only one part of a transnational effort to enhance the capabilities of Beirut's traffic police through training and the provision of equipment. Colonel Mohammad Al-Ayoubi, the director of the traffic police, described to me in summer 2010 how an agreement between the United States and the ISF endowed his division with new vehicles—mainly new Dodge Chargers and Harley Davidsons—as well as parts needed for older Harley Davidsons.[10] This delivery was part of the stated goal of the U.S. State Department's Bureau of International Narcotics and Law Enforcement Affairs (2009) to "assist legitimate, professional law enforcement institutions in Lebanon and their effort to protect Lebanon's territory and sovereignty." Al-Ayoubi told me of another recently arrived import: consultants from the French Ministry of the Interior had begun training Lebanese in the areas of traffic-management technologies, on-the-ground traffic control at intersections, and intrapolice communications.

This multinational brigade of development assistance dedicated to traffic safety reveals something about the country's geopolitical significance and the interest of foreign powers including, but not limited to, the United States, Iran, the Arab Gulf, and the European Union, in establishing and maintaining their spheres of influence. It is also related to Lebanon's chief economic source, tourism, and the need for managing a public and infrastructural realm, particularly in and around Beirut, whose summer population can swell with nearly two million visitors in a country of about four million.[11]

Traffic management was under the jurisdiction of Colonel Al-Ayoubi, and he expressed disappointment with the status quo. For him, driver behavior, not traffic management, was the problem. In the midst of Interior Minister Baroud's intensified efforts to enforce traffic laws, Al-Ayoubi explained: "Right now [summer 2010] we are making a campaign about seatbelts; we are stopping people and ticketing them for not wearing their seatbelt, and so, for right now, most will wear them. But if we stop, if we stop for one week, they will stop wearing them, and if you ask them why, they will say, 'Well, because the ISF is not making a campaign!'" I did not add, but considered doing so while he was talking, how I had recently witnessed one service taxi driver warn another about one seat-belt checkpoint up ahead in the road so the driver could put his on in time.

I met with Al-Ayoubi on a couple of occasions in his well-appointed office, replete with sofa, coffee table, and exercise bike, in an otherwise worn building housing the traffic police division of the ISF in southeastern Beirut. When we first met, in 2006, he held the inferior rank of captain. His office seemed to have been the only air-conditioned one in a dilapidated French Mandate-era building. He was stressed at work, receiving and answering frequent calls on his two-way radio. Four years later, he was more relaxed, though uninterested in discussing his promotion, and when I arrived at his office, I would find him in uniform but sitting and chatting with men in plainclothes.

A well-traveled and personable middle-aged man who talked of visiting his sister each year in the Washington, D.C., suburbs, he used the French word *mentalité* to describe what he saw as one of the primary challenges the police face: "There is a mentalité here, there is no respect for the law, ... and it's not because I am a man of the law that I say this. My response [to a comment about the ISF not being in the streets to enforce the seat-belt law] is this: In the United States, in France, is there a police officer at every intersection? Is it necessary to have one police officer for each car and driver?"

While traffic police cited drivers' irreverence for rules as a source of frustration, they also talked about the challenges posed by the social dynamics that pervade the police-citizen encounter. "In a country where the whole country knows the whole country," as one policeman put it, or "where if you don't yourself know someone you can call who can get you out of a ticket, you likely know someone who does," as another policeman said, the work of the traffic police involves daily streetside power struggles to apply the law.

For example, Al-Ayoubi spoke about how the police who work in the street, referred to by the military term *'anaser* (basic forces), get nervous about pulling over drivers of nice cars: "They feel like, how can I give this person a ticket? They might be someone important!" Two men who worked for Al-Ayoubi as supervisors and who together had more than twenty-five years of experience in the field concurred with this representation of how a routine traffic stop becomes a more complicated scene of status jockeying that necessitates their intervention. As self-described tongue-in-cheek "diplomats," they are called to scenes like these, scenes characterized by class- and status-based challenges—"Do you know who I am?" or "Who are *you* [to try to exert power over me]?"—born from the policeman's attempt to constrain certain drivers' mobility. This same kind of discursive disruption of the universal and fair application of the law—effected through the use of the phrase "Do you know who you're talking to?"—is described by DaMatta (1991) as a ritualized urban practice in Brazil that is emblematic of its uneven citizenship.

In this way, the traffic police, though endowed with power vested by the state, anticipate situations in which their power could easily be undermined by the workings of class and status. They occupy a vexed position in the margins of the state in part because their ability to enforce the law is disrupted by and contingent on localized relations of social hierarchy. These are relations that might be worked out right there on the street, when positions of more and less power are made legible through verbal interaction, or, as in the example Al-Ayoubi gave, they create problems for the traffic police even in their anticipation.

The urgency of these confrontations between the status-wielding driver and the traffic policeman is perhaps best related by the fact that they were described by the two supervisors as being on a par—that is, they required the supervisors' presence—with accidents "in which someone is wounded." Because they hold and display, through their uniform, the rank of officer, the supervisors said, they are able to help work out problems like these that might arise between a policeman and a driver. As they arrive to the rescue of a policeman under siege streetside, astride the traffic-police division's newest fleet of Harley Davidsons, the officers symbolize the vertical authority and disciplinary power of the state but also the effort that goes into enacting statist authority. At the same time, however, by signifying the verticalization of state authority,[12] the officers undercut the power of its lowest ranking members: if it is understood that the first policeman that pulls you over is powerless to enforce the law, that it is only his superiors you need to engage with, why not ignore the first policeman and hang around for the next?

Moreover, in seeking to "show the force [of the state]," which is how the officers described their actions, we see in these encounters an example of "modern statecraft being frustrated in its goal to reduce the chaotic, disorderly and constantly changing social reality beneath it to something more closely resembling the administrative grid of its operations" (Scott 1998, 82). The dynamics of class and status, which are often expressed through a demonstration of having connections, as I discuss further below, thwart the traffic police's plan for operations, a key part of which is getting drivers to comply with their wishes. It is these kinds of interactions, and frustrations, that make the traffic police perhaps unlikely bedfellows with the drivers who complain about the police's ineffective management of traffic and, as well, the ill-functioning state.

One of the supervising officers waxed hopeful toward the end of our meeting about the potential for mechanization to mitigate the wrangling with drivers that is a feature of their job. When computers inside the police vehicle issue tickets remotely, he thought, and not through a direct encounter between and police and citizens, things will be easier. This understanding of how traffic management in Lebanon could be improved by putting the task in the hands of machines rather than people was one shared by Colonel Al-Ayoubi when he talked about changing how tickets are issued: "Now they write it [the ticket] by hand, but we have a project to computerize tickets so that they will be issued without contact between police and citizens. We are doing what we can to reduce contact between police and citizens."

Curiously, I did not hear talk of what might happen to all the traffic policemen if their job were to be performed by computers. Rather, the more pressing concern seemed to be devising a system that would take the human element out of traffic-law enforcement, not for convenience's sake, to reduce error, or to cut personnel costs, but to improve its chances by removing from it the personal, face-to-face encounter. "Reducing contact between police and citizens," as Al-Ayoubi put it, was a move toward depersonalizing the state-citizen interaction by withdrawing its agents from actual social contexts in order to avoid the activation of personal networks and hierarchical dimensions of class that can and do arise from these contexts. We can understand this reduction of contact between state and citizen, exemplified too by the CDR's installation of the city's first parking meters (as described in chapter 5), as an effort to abstract the state by reclaiming the boundary between a faceless, orderly state and the messy world of society. Enacting this boundary is a move to endow the state with a seeming coherence, power, and effectiveness, what political scientist Timothy Mitchell terms "the state effect" (1991).

AT THE INTERSECTION

For the traffic police, encounters with drivers can be fraught with unease about the dynamics of power set in motion by a routine stop, but bodily stress and strain are also part of the job. Describing his job's challenges, one policeman simply said, "And then there is the exhaustion. There is so much exhaustion." Shifts directing traffic at intersections are seven hours each, either from 6:30 A.M. to 1:30 P.M. or 1:30 P.M. to 8:30 P.M., and supervisors come to the intersections to relieve soldiers for half-hour breaks and shorter bathroom and water breaks. I had observed police directing traffic throughout the city, usually as a passenger in a taxi or bus, and, particularly at the thorniest intersections, it was impossible not to notice the difficulties they faced:[13] they visibly struggled, under the hot sun, to control mobility as drivers whose lanes of traffic a policeman had called to a stop crept into the intersection and darted across. This transgression would encourage others to follow suit. The social vulnerability of the police that emerged in their interactions with civilians who asserted class- and status-based rights to be above the law was joined by another type of vulnerability rooted in the physical demands of the job and the public display of their inability to bring drivers under their thumb.

Sodeco, an intersection named for the retail and business complex it abuts, was identified as one of the five worst in the city by the supervising officers. There, I found a shady seat along on a concrete barrier wall, a spot from which to observe at peak times in the late afternoon. Sitting and watching, I empathized with the policeman. One afternoon in particular stood out. The day was hot, as all Beirut summer afternoons are, and the policeman exhorted drivers to stop and go. These exhortations involved standard arm and hand signals but also yelling and shouting and, later, full body contact. After he signaled one lane of traffic to stop so another could start, one car from the lane that he stopped began to sneak into the intersection hoping to make a just-in-time crossing. The policeman, his back to the car, turned in time and began yelling at the driver and gesturing for the car to move backward out of the intersection. The car showed no sign of retreat. The policeman leaned over and extended his arms onto the hood of car and, as it reversed, appeared to push it back into position.

What struck me most about this incident was how the policeman, although he was eventually successful in constraining the vehicle's movement, looked silly doing so. The driver's behavior made the policeman, whose work standing in the street necessitates public exposure of the most literal kind, seem even more, well, exposed. I thought about how the traffic police serve as foot soldiers of the state in the sense that the public terrain of the streets makes up the front lines of state-making; they are a territory that, while marked by the belongings of various groups and interests, is used by all. The traffic police seek to establish order over this terrain, but, routinely, their authority finds little traction among drivers who, as one board member of YASA put it, "already have someone on the phone who can get them out of the ticket by the time the policeman gets to their window."

In describing their efforts to control traffic, the policemen reveal how their role as agents of the state, albeit lowly ones, is enmeshed with and often subordinated to the power of civilians. This power, derived primarily from class and status but also from a kind of irreverence toward what seems to be viewed as a toothless arm of the state, poses as much of an occupational hazard for the traffic police as the bodily stress and strain of their very physical work. A certain insecurity emerges from what we might otherwise understand to be their seemingly routine encounters with drivers; higher-ups in the bureaucratic structure recognize and try to address this insecurity through various modes of intervention including the pulling of rank to match or overcome drivers' deployment of status with that of the state's as well as techniques of mechanization that abstract the state and remove it from the realm of the personal and the social altogether. In their embodied and often futile efforts to manage the mobility of a restive, and socially differentiated, lot, the traffic police experience a kind of vulnerability that resonates with that expressed in civilians' claims that "there is no state."

During a bus ride I took in outer eastern Beirut, sentiments about how traffic problems signified not just an ineffective but also corrupt state were made plain. A car trying to turn around in the road blocked traffic, and the bus could not move forward. The bus driver, soon joined by several passengers, began yelling at the driver of the car. The woman sitting in front of me cursed audibly, using a common and explicit profanity to condemn not the driver, but driving in Lebanon. Infuriated, the bus driver put his hands up and called out to all of us beseechingly, "How can these people get driver's licenses? How can they drive? They don't know how to drive!" The woman in front of me spoke up again, issuing a succinct and stony reply of *wasta*.

Wasta refers to connections people have and use in order to obtain various kinds of benefits for themselves or someone else. These benefits range from the more short-term, such as expediting bureaucratic processes, to the longer-term, like finding a job or getting a promotion. As a type of favoritism based primarily on kinship ties, wasta resembles other networks of influence and patronage relied on by individuals in matters ranging from the momentous and potentially life-changing to the trivial and everyday.[14] In Lebanon, where state instability and civil conflict have devolved from a political structure based on the precarious balance of power among representatives of the nation's sectarian groups, wasta is said to thrive because of the expectation that needs will be met more quickly and effectively through the exercise of personal relations rather than through "official" and "just" channels of government, merit, and performance. While the literature on wasta emphasizes its effects on the business climate, possibilities for career advancement, and nations' economic competitiveness,[15] I found that for Beirutis, wasta was a language for talking not only about the state but also about class: the unconnected are disadvantaged, while those connected to well-established families and groups do better in life.

In brandishing the word *wasta*, the infuriated bus passenger demonstrated the rawness and immediacy of sentiments, expressed by ordinary residents as well as by members of the traffic police, that link vehicular mobility with key aspects of civic and political culture. In the context of getting around the city, talk of wasta conveys a critique of the state's culture of corruption, evidenced by the ability of certain drivers to buy licenses,[16] as the bus passenger complained, or to get out of tickets, as was suggested by the traffic police. At the same time, as a discourse from below, *wasta*, rather than the wider-ranging *fasad* (corruption), zeroes in on the fact that social inequality shapes the nature of one's everyday encounters with the state. While I usually heard Beirutis talk about wasta with frustration and disdain, as a form of class injury that serves as a hindrance to their social mobility and sense of civic agency, for a few, like the teenager at a driver-education school who gleefully waved her laminated card toward me telling nonchalantly of how her father used wasta to get her a driver's license, the wielding of wasta brought fulfillment.

Wasta is commonly talked about as something one either possesses or uses; often, in the beginning, middle, or end of stories Beirutis told about being frustrated with their dealings with the state, they would refer to their lack of it. This was the case when Kamal, a twenty-four-year-old whom I met volunteering at an event for a youth organization, told me a story about how after going to the police station to retrieve his towed car, he found it damaged and the fuel tank empty. He was sure that the car had been damaged and siphoned of gas while in police possession, but, he said, "It's not like I have wasta; . . . there was nothing I could do about it." He questioned the police working in the office but backed off quickly when no one offered help, adding, "I always avoid too much of a confrontation with the police." Kamal's sense that there was no recourse in his situation and, moreover, that in pressing the point he would only make himself vulnerable to further mistreatment was also evident in streetside scenes between police and residents. Although dealings with traffic police working at intersections might be less anxious for residents than those with standard police (the ISF's gendarmerie), for male drivers in particular,[17] those from ordinary backgrounds who do not drive cars bearing, for example, a high-status three- or four-digit license plate number (rather than the standard six), which retail for several thousand U.S. dollars,[18] being pulled over by any member of the police while driving is an interaction one tries to keep nonconfrontational.

Indeed, beyond a driver's mode of transport and make and quality of the vehicle, license plates are a way in which social inequalities are made public and are directly apprehensible to all. Those who can afford to rent cars, usually foreigners or returning overseas Lebanese, usually drive vehicles bearing green plates, whereas taxis have red ones, judges' plates display an emblem symbolizing the scales of justice, and, moving up the hierarchy, members of parliament—and sometimes their kin—have blue license plates, while those driving without any plates at all are at the highest echelon of power. This blatant hierarchy of license plates is part of the traffic policeman's calculations about drivers' status and his anticipation about the wasta these drivers may possess. Anxieties policemen face about pulling someone over who might be important are also born from being fish out of water in their encounters with cosmopolitan Beirutis as, according to the traffic-police supervisors I spoke with, many of the men working in the street "are not from Beirut; they are from villages, and here they are now living in dormitories, without their families [for the first time]."

For both police and civilians, the traffic encounter is a locus for everyday practices of corruption and social injustice that give rise to feelings of being frustrated by and disillusioned with the state. This frustration and disillusionment, I suggest, are part of the conditions of insecurity that surround life in a militarized urban landscape. The insecurity that shapes and is shaped by mobility in Beirut is a piece with concerns about political instability and the state's not providing a kind of safety net, social and otherwise, for its citizens.

THE STATE THAT COULD BE

I began this chapter by relating how Interior Minister Baroud, in his alignment with the traffic police in particular and crackdown on illegality more generally, openly shared citizens' sentiments about an ineffective and corrupt state. Within a conflict-laden and business-as-usual political climate, Baroud, by playing the part of a lone protagonist standing up to crime—even that of the powerful—seemed to represent the kind of hoped-for alternative latent in comments made by both police and civilians that "there is no state." Baroud's own trajectory illustrates not the impossibility of this alternative, that of a fair and just state that provides for and protects even its most ordinary citizens, but certainly the kinds of conditions challenging its realization.

In May 2011, Baroud resigned from his position after a final confrontation with General Rifi, director of the ISF. During a bizarre media-covered power play, Rifi, Baroud's subordinate, publicly flouted Baroud's orders to resolve a conflict Rifi had escalated with another government official. In his resignation speech, a dejected Baroud, who came off as a naïve idealist-reformer beaten down by the intricacies and divisiveness of Lebanese politics, spoke about the breakdown of the Lebanese state. In his final hour as a member of government, Baroud once again identified himself with regular folk fed up with the status quo.

Baroud's resignation illustrates one of the key findings of my research: everyday interactions with the state in their traffic encounters leave Beirut's citizens discontented and with unfulfilled aspirations for the well-functioning, respectable, and protective state that could be. His departure signaled the ways in which people, sometimes even those at the helm of governance like Baroud or the would-be state-makers such as the traffic police, maintain unmet expectations about a state that could, and should, work better. In this sense of disaffection with the state, which is an outcome of unfulfilled desires for, in part, security and safety, we see a different kind of retreat of the state than that which we tie to the ushering in of neoliberal governmentality. This retreat is marked by the precedence of players participating in high-stakes games of political conflict and negotiation over concerns for the common good. In a space where people say there is no state, the traffic encounter in fact represents a crucial site of state-making. In a city wracked by conflict, in a militarized urban landscape territorialized by divisive political claims, the policing of traffic might also be understood as a seam that endeavors to fasten together, under the domain of state authority, civic space itself.

The encounter of the traffic police and driver provides an ethnographic view of bureaucratic embodiment in public space, one that illustrates the kinds of mundane practices that go into the project of state-making. It shows us too the multiplicity of actors who, as Abrams phrased it ([1977] 1988), make up what a "state" is thought to be.

CONCLUSION

In June 2013, I was riding in a service taxi driving along the Beirut highway that heads north from the airport area to downtown. It is an elevated highway that affords passengers a panoramic view of building rooftops. On one rooftop, just off to the side of the highway was a distressing sign of the times: swaying in the light breeze was the black flag of the al-Nusra Front, an Al Qaeda affiliate fighting in Syria against the Assad regime.

In summer 2013, the impact of the Syrian civil war was visible throughout the city. The sight of the al-Nusra Front flag, for example, demonstrated the war's splintering effect on Lebanon's already jagged political and security landscape, as supporters of groups fighting the Syrian regime clashed with those, like Hizbullah, who backed Assad's forces. Hizbullah's leader Hassan Nasrallah had just recently made the public announcement that Hizbullah soldiers were fighting alongside the Syrian army in order to defeat rebel groups who controlled areas that bordered Lebanon.[1] Never before had Hizbullah guerrillas waged war outside Lebanon. These border areas in the east, the sites of rocket-fire exchange between armed groups, became just one of the many violent hotspots in the country that challenged the state security apparatus as militarized factions who supported the Syrian rebels in the northern city of Tripoli and an Islamic extremist group in the southern city of Saida also carried out regular attacks on Lebanese army deployments.

A deteriorating security situation and deepening political divides were not the only effects of the war in neighboring Syria, as the arrival of what is estimated to be, at the time of this writing, more than a million refugees strained Lebanon's resources.[2] One service driver put it this way: "The

Syrian war is burning Lebanon." This was his response to my question *kayf al-wada'a?*—literally, "How is the situation?" but really more like "How are things going?" All these Syrians who have come into Beirut, he went on to say, "they are coming in and taking our jobs; they start driving taxis that are not registered, and now there are too many taxis on the road. We can't survive like this." During summer 2013, I heard many comments like this one about how people were being squeezed by the war. While the majority of Syrian refugees were living in settlements and villages in other parts of Lebanon, many Syrians, from a range of class backgrounds, had come to Beirut and, for the first time, not as male laborers on their own, but as families. Beirutis remarked on how the presence of Syrians was contributing to climbing rents, to increased competition for jobs,[3] and to a palpable sense of crowdedness in the very few free public spaces that city residents could call their own. On the streets of the city, Syrian license plates on private cars from areas roiled by war—Homs, Hama, Aleppo, Damascus—were a common sight. In this sense, the Syrian war and catastrophic humanitarian situation it has produced have placed both Syrians and Lebanese who live and work in Beirut and elsewhere in the country in an increasingly insecure position as they have had to forge lives amid a contentious and violent political-economic landscape.

MOBILITY AND THE UNEVENNESS OF CITIZENSHIP

What struck me about the service driver's plaintive statement—"we can't survive like this"—was how it recalled stories about experiences of mobility that I had heard throughout my research, stories that form the core of this book. For my respondents, getting around Beirut meant having to confront and cope with the effects and conditions of a volatile geopolitics that is both domestic and regional. In summer 2013, the mobility situation was plainly evident, as traffic congestion intensified as a result of the influx of cars owned by the city's new Syrian residents and of sudden road closures established in response to the tides of political events and activities. One quiet Sunday morning in June, for instance, on my way from the east to the west side of the city, I encountered an army blockade surrounding the downtown area; it had been set up in anticipation of demonstrations against Hizbullah's participation in the Syrian civil war. These kinds of sudden road closures not only changed routes of travel, they instantiated in public space the proximity of danger and the possibility of urban warfare. For the service driver, whose livelihood depends on the rate at which he makes his way through the city streets, these traffic problems posed another kind of threat, as they raised the specter of the inability to make ends meet. In these different ways, as my ethnography has shown, the field of mobility is central to the experience of urban citizenship in Beirut.

So, what does mobility tell us about what it means to be a citizen in this place? For one, it tells us something about the role inequalities in spatial access and movement play in producing an uneven citizenship. In my research, I found that residents of the city understood movement through public space as a context in which class and status were thickly written.[4] The privileged and the well-connected were understood to be able to buy off police, for example, and the secured—and sometimes itinerant—enclaves of VIPs intensified processes of "elitization" in the city by stripping away, sidewalk by sidewalk, more public space from the public. The realm of mobility, in this sense, exposed some of the inequalities that constituted the city itself. Mobility was also an arena through which the workings of political sectarianism were produced as the city's political sectarian geography shaped, and was also shaped by, experiences and understandings of spatial movement. Feelings of safety and unsafety that had to do with the city's political sectarian territorialization—a territorialization marked by a multilayered temporality of conflict comprising the past, the present, and the anticipated future—emerged in stories people told about which routes they and their children would and would not take. Mobility, as I have illustrated, played a critical role in differentiating residents from one another not just by class and status but also by political affiliation.

By focusing on the ways in which mobility and spatial access produce an uneven urban citizenship in Beirut, I have taken a different path from that traversed by much of the scholarship on the city's social and political geography, which emphasizes how sectarianism has configured, divided, and otherwise shaped the experience of space and place not just in Beirut but in Lebanon as a whole. Without question, sectarian difference is vital to any understanding of the social constitution of space in Beirut, and yet, as I have shown, it offers us only one lens on this process. By providing insight into the ways in which class and status configure how people live and move through the city, I have aimed to contribute to studies of the Middle East that address the significance of social class in urban life through a perspective that goes beyond a focus on consumption practices and ideas of modernity and draws our attention instead to the ways in which class and status are enacted in the civic and public realm of the streets.

ENCOUNTERING THE STATE

The stories about and experiences of moving through the city I have traced in this book also tell us something about the relationship between citizens and the state. In my conversations with Beirutis about traffic and mobility, the abstract category of "the state" was used by people in everyday parlance, but there was also a particular idea of the state, as an entity of government endowed with the responsibility of securing its citizens in the public realm, that emerged as salient. A lack of state regulation of the chaos on the roads led to an excessive number of accidents and injuries, many pointed out, and practices of going around and above the law—purchasing driver's licenses, for example—have riddled the state with corruption. In short, Beirutis framed the lack of road safety and the arena of mobility more generally through references to the state. In this way, in my research, I found support for anthropologist Begoña Aretxaga's argument (2003, 395) about the staying power of the state. While processes of neoliberal capitalist globalization have eroded many functions of the state, she argued, the state remains a crucial social and political presence. In Beirut, everyday mobility, as I have demonstrated in this book, was an important discursive context not only through which people registered civic concerns about the state that is but also through which the resources and powers of the state that could be were imagined.

Notions of being unprotected by the state also emerged in the ways both civilians and the police described the traffic situation using the phrase *ma fi dowla* (there is no state). This expression of statelessness conveyed an unmistakable cynicism about the state. But something more than cynicism framed the structure of feeling toward the Lebanese state among my respondents. Amid, and also as a consequence of, an unstable and insecure postwar climate characterized by the devastation of war with Israel in summer 2006, ongoing electricity shortages, the continual sparring of transnationally backed internal political groups, and the flowing over of conflict from the Syrian war, I heard in the cynical remarks about there not being a state aspirations for one that worked differently and that offered security and protection for its citizens. The ill-functioning state that citizens did not feel protected by on the roads is the same one that could not claim or even pretend to protect city dwellers from other kinds of vulnerabilities: the threat of regional war, for instance, as well as economic and health-related insecurities associated with soaring rents, a limited social safety net, and a lack of environmental policies aimed at reducing increasing rates of pollution (Fawaz, Harb, and Gharbieh, 2012, 180).[5]

Through my discussion of spatial mobility as a social and discursive register through which Beirutis map out their experiences and understandings of the state, I have sought to take seriously people's sentiments about feeling unprotected by the state. In doing so, I have made conceptions about the weakness of the Lebanese state the basis of my analysis rather than the concluding point, which is often the case in academic and policy discourse that relies on the paradigm of the "weak," "failed," or "fragile" state in the examination of Lebanon. In light of Philip Abrams's call to demystify the idea of the state,[6] my ethnography of the embodied labor of the traffic police shows the state in its nitty-gritty details as a set of practices and efforts that are undertaken by a multiplicity of actors in everyday ways. The perspective of the state I have provided is thus one that considers state governance to be a multifaceted human and social enterprise instead of a coherent and autonomous entity.[7] This is a public and spatial enterprise comprised, for example, of traffic policemen at intersections, patrols set up to catch helmetless motor-scooter drivers, and, sometimes, the direct on-the-ground actions of leaders like Interior Minister Ziad Baroud, who once physically removed illegal barriers that impeded traffic flow in the downtown area.

MOVING THROUGH THE INSECURE CITY

Theorist of space Henri Lefebvre conceived of the city as the site of encounter par excellence, a place where different kinds of people not only interact but where certain kinds of collectivities might be formed.[8] As my ethnography has shown, the experience of being mobile in the city is central to this encounter. In their mobility experiences, drivers, walkers, and passengers in Beirut navigated the city's class, political, and militarized geography through encounters with traffic police, roadside displays of party politics, and each other. These prosaic scenes of getting around the city, I have argued, are in fact sites of uncommon civic significance that produce—and in turn are produced by—urban citizenship in a place where and a time when everyday life is framed by intersecting forms of insecurity. By focusing on the physical movement of people through urban space, I have provided a departure from anthropology's long-established concern with human movement across transnational borders, and I have engaged public scenes of spatial mobility, from rides in cars with student drivers to pedestrian navigations around security barriers and conversations between passengers in shared taxis, to demonstrate how class, politics, and state power are spatialized in the urban public realm.

During my research, I found the constellation of experiences that constitute this urban public realm to be characterized by dynamism and effervescence, to be sure, but also by discontent. Pervasive corruption among the power holders, the private takeover of public space, and an inefficient state that leaves citizens to fend for themselves prevailed in the stories Beirutis told about their mobility experiences. In these stories, people expressed not just criticism of but also disaffection with a system of governance and geopolitical landscape that rendered them insecure. In these expressions of disaffection, captured by the kinds of comments service taxi drivers often shared with me—"Lebanon is run by just a few people" and "If they [the powers that be] want there to be a war, then there will be; they decide what will happen in Lebanon"—I heard the desire for an alternative civic and political order. Amid the fractious politics of the country and region, it remains to be seen how this alternative, and the lives of Beirut's urban citizens, might take shape.

But people's lives in Beirut are made insecure not only by public violence and Lebanon's contentious geopolitical environment but also by the challenge of making ends meet and of coping with concerns about downward social mobility and the lack of a safety net. Thus, this ethnography is also about lives that are part of the broader global experience having to do with the anxieties about being unprotected that characterize the human condition in the early twenty-first century. Moreover, the militarized and secured urban setting that Beirutis navigate on a day-to-day basis is increasingly salient for the rest of the world, as the intensification of policing, surveillance, and security are quickly becoming one of the central features of life in the contemporary city. The experience of being mobile in Beirut, then, tells us about the present in a zone of conflict and also, potentially, about our future.

NOTES

INTRODUCTION

1. Hizbullah, "the party of God," is an Islamic resistance group formed after the Israeli invasion and the arrival of the U.S. military in 1982, during the civil and regional war (1975–1990). The group, which I discuss further in chapters 2 and 3, was founded to defend Lebanon against and to liberate the country from Israel occupation and Western control. For more on Hizbullah's history and growth, see Norton (2014).
2. "Israel/Lebanon: Israeli Indiscriminate Attacks Killed Most Civilians," Human Rights Watch, accessed September 2, 2008, http://www.hrw.org/english/docs/2007/09/06/isrlpa16781.htm.
3. For recent anthropological work that considers the role of mobility in producing urban inequality, see Caldeira (2012) and Lutz (2014).
4. Services are shared taxis flagged down in the streets. They are the most widely used form of public transit in Beirut. During my first research period, September 2004–June 2006, a ride in a service cost 1,000LL (about US$0.66). Following the war with Israel during summer 2006, according to Ilham Khabbaz, whom I met with in summer 2010, of the Ministry of Public Works and Transport, the Ministry agreed to the request of the Taxi Drivers Syndicate for an official fare increase of 500LL. In practice, she said, drivers asked for 2,000LL rather than 1,500LL and received this amount from passengers. Thus, following the 2006 war, the fare for a service ride became 2,000LL (about US$1.30)
5. Popular and cultural productions focused on the city's traffic situation include the *Al-Fassad* (Corruption) television news show on July 9, 2010, with guest Ziad Aql, director of the Youth Association for Social Awareness (YASA), a traffic-safety organization; the theatrical production *'Ajat as-sayr* (Traffic Jam), staged at the Monnot Theater in Beirut in 2013; the documentaries *Taxi Talk* (2009) and *Taxi Beirut* (2011); and the popular web-based video series *Shankaboot*, which chronicles the life of a delivery worker on a motor scooter (see http://www.shankaboot.com/).
6. Often described as a civil war, the protracted conflict in Lebanon (1975–1990) had both domestic and regional dimensions and was fought by the armies of several nations. See chapter 2 for a complete overview of the war.
7. In particular, I draw on Low's (2004) and Caldeira's (2000) ethnographic investigations of sociospatial inequality in my theorization of how social power is constituted through spatial movement. I share Low's and Caldeira's conceptual approach to space: that space is socially constructed and that the social is spatially constructed as opposed to the idea that space is a mere container or "setting" for social life.
8. The understanding of sectarian identity that I hope to convey here is a nonessentialist one that rejects the idea of there being a kind of necessary correspondence

(Hall 1985) between sectarian belonging and other characteristics such as political ideology, religious belief and practice, and class status. Indeed, features of identity do not match up in such an uncomplicated way in any context, and Lebanon is no exception. Many Lebanese reject the idea that they can and should be defined in sectarian terms, and this sentiment is formalized in the vibrant activism of antisectarian civil society groups whose aims range from the establishment of a unified and secular (rather than sect-based) civil code that governs personal-status laws to the toppling of the "sectarian regime." And yet, because Lebanese religious identities have been mobilized for political and social purposes since the nineteenth century (Makdisi 2000), sect remains a meaningful indigenous category. In this vein I emphasize the political aspects of sectarian identity in this chapter, rather than those that have to do with religious belief and practice, in order to investigate the role political sectarianism plays in configuring everyday geographies. Thus, I use the term *political sectarianism* primarily to highlight the ways in which sectarian identity is animated through mechanisms of the state and citizenship that are governed by and through the workings of party politics.

9. Deeb and Harb (2013, 234n17) note that "prior prisoner exchanges between Hizbullah and Israel set the precedent for Hizbullah's capture of Israeli soldiers to use in future exchanges." For more on the July 2006 war, see "The Sixth War" (2006).

10. These studies include Ghannam's (2002) on Cairo and Kanna's (2011) on Dubai.

11. Here I draw on the notion of the wounded city as explored in Schneider and Susser (2003).

12. For ethnographic examinations of spatial movement amid conflict and insecurity in this region, see Ochs (2011) on Israel and Allen (2008) and Peteet (forthcoming) on Palestine.

13. See, for example, Amar (2013) on governance by the security state in Cairo and Rio de Janeiro and Penglase (2014) on the daily insecurities of life in a Brazilian favela.

14. For an investigation of what it looks like on the ground when citizens move to protect themselves and their communities, see Goldstein (2012).

15. See Gusterson and Besteman (2010) for a multifaceted approach to the kinds of lived insecurities faced by Americans in the face of growing inequality and weakened structures of social support in the United States.

16. On the experience of checkpoints and closure for Palestinians, see Allen (2008); Hammami (2004); Peteet (forthcoming); Wick (2011).

17. In a social and psychological analysis of emotions in everyday situations, Katz (1999) closely examines the emotions of driving.

18. A group of scholars (mainly from Europe and the United Kingdom in the fields of sociology and human geography) have forged an interdisciplinary field of mobility studies (see Hannam, Sheller, and Urry 2006), which investigates the physical movement of people and objects as well as the movement of images and information across space. These mobility studies have drawn our attention to the broader social processes that produce and inhibit movement but, in their abstract and case-study approaches, have been less effective in providing concrete insights into the different kinds of power relations that surround everyday mobility practice. There is also a significant body

of historical work on automobility that focuses primarily on the United States and Europe.

19. The contributions of feminist thought to our understanding of space and place have been interdisciplinary and include, to highlight just a few areas of research, studies of housing and interior design (Hayden 1982), gated communities (Low 2004), and community organization (Pellow 2008), as well as theoretical projects that integrate gender analyses into critical geography studies of capitalism and economic restructuring (Massey 1994).

20. Fenster and Hamdan-Saliba (2013) highlight these issues in their review of gender and feminist geography in the Middle East. Works that take up issues of gender and space in Middle Eastern cities include Ghannam's (2013) research about young men and women in low-income Cairo, which traces gendered experiences of public mobility; Newcomb's (2008) account of gender and identity in Fes, Morocco; de Koning's (2009) ethnography of elite consumption practices and uses of Cairo's public space; and an edited volume by Rieker and Ali (2008).

21. Ethnographic works concerned with masculinity in the Middle East include Ghannam (2013) and Inhorn (2012).

22. Here I am inspired by theorist of space Henri Lefebvre ([1968] 1996), who linked enactments of citizenship to the urban through what he referred to as "the right to the city"—that is, the right to inhabit the city in the broadest sense and thereby actively produce the city. Relations of power confound exercise of this right as users of the city confront limits to their spatial, civic, and political possibilities by structures of dominance and control.

23. Increasingly, scholars have seen modern citizenship not only as a category of legal and political membership in a nation-state but also as a site of "multiple logics of belonging" (Vora 2013, 33) informed by both historical and contemporary cultural, social, and political economic experiences of, among others, transnationalism, migration, diaspora, and consumerism.

24. This literature confounds the state-society binary in various ways by, for example, interrogating theories of the state that endow it with a coherence, autonomy, and unity that it does not possess (Mitchell 1991), conceiving of the state as being made up of a multiplicity of actors (Abrams [1977] 1988), examining the spatial characteristics of the state (Gupta and Ferguson 2002), and undertaking ethnographic investigations of state-citizen interaction (Sharma and Gupta 2006).

25. See chapter 6 for more on this understanding of Lebanon as a weak state.

CHAPTER 1: THE PRIVATIZED CITY

1. The first Palestinian refugees settled in and around Beirut after the establishment of Israel in 1948 and significantly more arrived following the 1967 Arab-Israeli war. The two largest refugee camps near Beirut—Burj al-Barajneh and Sabra-Shatila—are adjacent to the Hizbullah-controlled neighborhoods of the southern suburbs. The United Nations Relief and Works Agency for Palestine Refugees in the Near East (UNRWA)

reports that many, but not all, of the 455,000 Palestinian refugees registered in Lebanon live in camps (see UNRWA's webpage about Lebanon, http://www.unrwa.org/where-we-work/lebanon). Living conditions for the majority of Palestinians in Lebanon, both within and outside the camps, are characterized by poverty, disenfranchisement, and overcrowded habitation with inadequate infrastructure. Those living in the camps and informal settlements around Beirut experience, in addition, social and spatial exclusion effected through their geographical isolation from other parts of Beirut as well as constraints on access to the camps by nonresidents (Hanafi, Chaaban, and Seyfert 2012; Peteet 2005). In 2005, officially registered Palestinian refugees were granted permission by the Lebanese state to work in the clerical and administrative sectors for the first time, but they are still prohibited from seeking employment in many other professions and can neither legally own nor inherit property or petition to gain Lebanese citizenship (Khalili 2007, 56). Refugees also have extremely limited access to governmental facilities, including schools, and no access to public social services. Most rely on UNRWA and, increasingly, NGOs for education, health, and social services. The Lebanese government justifies the lack of Palestinian civil and political rights—and representation—by claiming that Palestinians are only temporarily residing in Lebanon. For more on Palestinians in Lebanon, see Allan (2013); Hanafi, Chaaban, and Seyfert (2012); Khalili (2007); and Peteet (2005).

2. This estimate from UN-Habitat (2012) includes neither migrant workers nor refugees who have come to Beirut since the escalation of the Syrian conflict in late 2011.

3. Lebanon has six administrative governorates that are divided into twenty-five districts and then subdivided into municipalities. Municipal Beirut comprises a governorate and a district whereas its suburbs are municipalities housed within other districts.

4. According to Fricke (2005, 177), nearly half of Beirutis rent their homes, and tenants typically pass their leases on through family members.

5. Beirut's downtown area is variously called in English the "Beirut Central District" (or BCD for short), which is how the area's postwar developers often refer to it; "Solidere," which refers precisely to the company that redeveloped it after the war; and "downtown"; in Arabic, *wast al-balad;* in French, *centre-ville* (city center or downtown).

6. Sociologist Samir Khalaf (1985, 231) discusses the role of private property in the planning and design of Beirut and also its significance as a source of wealth.

7. For more on the Armenian community in Lebanon, see Migliorino (2008).

8. Extending across the midsection of much of the country, the Lebanese mountains are a site of biblical importance.

9. Gates (1998) outlines how the development of Lebanon's "tertiary" (service) sector became the most productive segment of the economy and how the mercantile elite—in collaboration with French commercial interests—succeeded in creating an extreme state of laissez-faire.

10. The *Tanzimat* reforms created conditions favorable for investment in such projects as the construction of a toll road between Beirut and Damascus and the rebuilding of the city's water-supply system (Fawaz 1983, 78).

11. Religious violence of the nineteenth century, which culminated in sectarian mobilizations and massacres in 1860, was, as Makdisi (2000) argues, the outcome of a joint "modernizing" effort by Ottoman and European powers that reconfigured the basis of political loyalties from peasant-noble to sectarian ties. This new culture of sectarianism reconfigured the geography of Lebanon into distinct sectarian communities and "singled out religious affiliation as the defining public and political characteristic of a modern subject and citizen" (174).

12. Hanssen (2005) and Sehnaoui (2002) provide elaborate descriptions of these leisure sites.

13. At the outbreak of war, the Ottoman government abolished Lebanon's semiautonomous status and established a military occupation of the country. Anti-Ottoman political activity was violently repressed (Volk 2010), and food shortages during the war—aggravated by a locust plague in 1915—which were an outcome of shipping interruptions and military requisition of supplies by the Ottoman army, resulted in a famine that claimed the lives of more than one hundred thousand residents of Beirut and the Mount Lebanon region (Traboulsi 2007, 72).

14. Maronites are an Eastern Catholic community that takes its name from a fourth-century monk. Since the sixteenth century, French Catholic missionaries had developed ties to the Ottoman Empire's indigenous Catholics and had seen themselves as their protectors (Dueck 2012).

15. Maronite political influence was expanded through their placement in positions in the high commissioner's administration in Beirut, for example, and their increased representation in elite military divisions (Picard 1996, 65; Salibi 1988, 35). These forms of patronage and centuries-old cultural and economic ties meant that Maronites were loyal to the newly established state whereas many Muslims who had aspired to create an independent and unified Syrian Arab nation constituted by its natural geographical borders rather than those drawn by the colonial power rejected the mandate.

16. Owen (1976) analyzes the underdevelopment of industry and agriculture as a direct consequence of the decision to create a Greater Lebanon, whereas Gates (1998) emphasizes this underdevelopment as being an outcome of the coalition of financial interests among the national mercantile elite.

17. Social Watch provides a concise description of the country's regional disparities; accessed December 17, 2014, http://www.socialwatch.org/node/10767.

18. Naiden and Harl (2009) argue that Beirut's commercial strength and thus its role as a political asset make it a veritable city-state.

19. See Thompson (2000) for analyses of French rule through paternalism and a mediating elite.

20. There are eighteen different sects documented as existing in Lebanon. While English-language texts sometimes use the words *communitarian* or *confessionalist* to describe Lebanon's political system, the Arabic words *ta'if* (sect) and *ta'ifi* (sectarian) were those used most often by my respondents to describe the political and social structure. For this reason, I adhere to the usage of these latter terms rather than the former except when citing others.

21. The French urban planner Baron de Haussmann transformed Paris in the 1860s. Features of this transformation included the creation of wide boulevards, architectural uniformity achieved through uninterrupted building façades, and the production of consumer and civic areas for social and economic use by the "modern" French bourgeoisie.

22. However, in the end, not all aspects of Haussmann's Paris could be exported. Two of the avenues that radiated out from the point of the "star" at Place de l'Etoile were cut off because they ran into three important historic religious structures (Davie 2003).

23. In 1922, Ottoman Turkish was abolished in schools where it was still taught and French and Arabic were made the official languages of Lebanon. Language, as historian Nadya Sbaiti notes, was central to how the mandate extended its discursive and physical authority: "Language was the means through which people could assert or express their own national sentiments, and, particularly in Beirut, it was also the axis around which national, religious and class affiliations were formulated by the residents of the new Lebanon" (Sbaiti 2009, 77). See Dueck (2010) also for a discussion of the role of education in French governance of Lebanon.

24. Saliba (2004) considers Beirut's eclectic mix of Western and Eastern architecture as an articulation of its "cultural dualism," which itself constitutes a kind of national style.

25. Constantinos Doxiadis (1958), the French development organization Institut International de Recherche et de Formation, Education Cultures Développement (1959–1964), Michel Ecochard (1961–1964), the Atelier Parisien d'Urbanisme (1977), and more recently the Institut d'Aménagement et d'Urbanisme de la Région Ile de France (1983–1986 and 1991 onward) have developed master plans for Beirut. For more on these master plans, see Verdeil (2005).

26. The planning system in Lebanon is complex. Planning occurs at three main levels: national, regional, and municipal. At the national level, the Directorate General for Urban Planning within the Ministry of Public Works and Transport develops planning regulations and master plans and issues building permits for some municipalities. The Council for Development and Reconstruction (CDR) is a semi-governmental agency linked with the Council of Ministers, which is responsible for the allocation of the majority of funds earmarked for the post–civil war reconstruction of Lebanon, supersedes all other public institutions in implementation decisions, and governs a body called the Higher Council for Urban Planning. The municipal and regional levels (regions are collaborations or federations of municipalities) are responsible within their geographic territories for planning and day-to-day maintenance of infrastructure, public transportation, and so forth. Some municipalities issue building and construction permits. For more on Lebanon's planning organization and processes, see Stevenson (2007). See UNDP (2011) on Beirut as haphazard, Khalaf and Kongstad (1973) on the city as unplanned, and Perry (2002) on Beirut as unregulated.

27. As an example of this kind of political infighting, it was widely reported that Ghazi Aridi, head of the Ministry of Public Works and Transport, was forced out of his position in 2013 as a result of his clashes with Walid Jumblatt, a fellow Druze (a minority religious sect in Lebanon), but a more powerful, political leader.

28. See chapter 3 for a discussion of Solidere's rebuilding of downtown. Solidere is an acronym for Société Libanaise pour le Développement et la Reconstruction de Beyrouth, French for the Lebanese Company for the Development and Reconstruction of Beirut.
29. For scholarship that examines these processes in the Arab world, see Elsheshtawy (2008); Singerman (2011); and Singerman and Amar (2006).
30. Solh made these comments at a public symposium, "City Debates," held at the American University of Beirut on May 12, 2006.
31. This issue of varying setbacks and lack of regulation of the built environment is discussed in a UNDP report (2011, 252).
32. Beirut's first street atlas, *Zawarib Beirut*, was first published in 2005 and was aimed primarily at an audience of Lebanese ex-pats returning to the city for leisure and business; they found it challenging to navigate the city without precise street names and addresses.
33. See Roy and AlSayyad (2003) for a discussion of informality in contemporary urbanization processes.
34. While Krijnen and Fawaz (2010) track the increase in these practices of informal decision making and allowing exceptions to the law, Fawaz (2009c) demonstrates the historical precedence for this "building by exception" through a discussion of planning regulations and illegal housing in Beirut's peripheries. According to Riachi (1963), until 1945 the prevailing type of construction in the city was the three-story, two-apartment walk-up building. Seven- and eight-story buildings appeared during a construction boom after 1945, and later, in 1954, a special amendment to the building code was passed by Parliament permitting a maximum of nine floors per building. Krijnen and Fawaz (2010) provide a detailed analysis of processes of informality in the planning, design, and construction of Beirut's built environment.
35. Zoning exists for only some of Beirut's land, and rezoning, after original regulations "expire," is common (Glasze and Alkhayyal 2002, 332). Khalaf has written about how informal patron-client relations influence zoning to the degree that "virtually everyone within the government civil bureaucracy—from the simple municipal clerk who overlooks a minor transgression to a high government official who intervenes on behalf of either his client or patron to reroute a road network or rezone a certain area—is placed in a strategic position to affect the redistribution of rewards and benefits. Such manipulations are especially frequent when the case involves land or real estate" (1985, 230).
36. Here I draw on Fawaz's (2009a) insights into the role of Hizbullah in urban planning and construction.
37. For more on energy hawking in Beirut, see Justin Salhani's article on the website for the nonprofit organization Next City, "Power Cuts in Beirut Spawn an Informal Energy-Hawking Industry," accessed December 17, 2014, http://nextcity.org/daily/entry/power-cuts-in-beirut-spawn-an-informal-energy-hawking-industry.
38. The French built the first railway in Lebanon at the end of the nineteenth century. For much of the twentieth century, there were lines connecting Beirut and Damascus, the eastern Bekaa Valley with Aleppo, and another that ran along the coast. The rail network, which had already fallen out of use by the 1970s because of the popularity of cars and buses, was destroyed during the civil and regional war and has not been revitalized.

39. A 1931 announcement in the Lebanese newspaper *La Syrie* about the founding of a new organization, the Association of Automobile Importers, championed the benefits of the automobile for the tourism sector and urged the government to ensure that the nation's roadways were improved so as to reap these benefits: "The vehicle, with its indispensable ally, the good road, encourages and facilitates the arrival of tourists to a country. . . . Lebanon, once the seat of humanity and of religions and civilizations, is sure to attract an increasing number of tourists provided that a wise legislation facilitates and develops the touristic effort" (*La Syrie*, May 8, 1931).

40. "Beirut Tourist Police Attempt to Silence Automobile Horns" (1964) details the roles and responsibilities of this special traffic-police force.

41. Nakkash and Jouzy (1973) offer a picture of the traffic congestion during this time.

42. These and other statistics regarding traffic in Beirut can be found in a report by Aoun et al. (2013).

43. According to Ilham Khabbaz, whom I met with in summer 2010, director of Land Transport at the Ministry of Public Works and Transport, intensifying traffic congestion was also an outcome of the growing number of public transportation vehicles operating illegally without proper licensing—including shared service taxis, minibuses, and buses.

44. The suffix *ayn* denotes the dual or two of something in Arabic. Lebanese commonly add this to non-Arabic words like *service* or *Bonjour*.

45. For more on the city's parking problems, see a 2011 UNDP report (244).

46. See, for example, Nahnoo (Arabic for "we"), a youth organization (nahnoo.org), and the Beirut Green Project (beirutgreenproject.wordpress.com).

47. For more on the city's pollution problem, see Chabban, Aoub, and Oulabi (2001) on air quality, Korfali and Massoud (2003) on noise, and a UNDP report (2011) on pollution and environmental degradation.

48. The Fouad Boutros Highway project, which was still in the proposal stage as of this writing, is a controversial plan to build a 1.3 km highway linking Ashrafieh with the port area. The project has been widely protested by residents who argue that it will destroy historic properties and increase neighborhood traffic.

49. In their work on the political economy of urban development, Logan and Molotch (1987) outline the political, planning, and investment processes that constitute the city as a "growth machine" in which the urban landscape is developed for its exchange rather than use value.

50. Sarkis has been an active in both public and scholarly forums about Beirut's urban development; see, for example, Rowe and Sarkis (1998).

51. The notion of the right to the city is an idea first proposed by Lefebvre ([1968] 1996). Lefebvre conceived of it as the "right to urban life," a right with which ordinary residents of the city are endowed and which can become a basis for class-based political and structural change. David Harvey (2008) reexamines the right to the city through the relations between urbanization and capitalism and calls for a democratization of the power to shape processes of urbanization.

CHAPTER 2: THE SPACE OF WAR

1. For more on the National Museum and its emphasis on a Phoenician national narrative that orients Lebanon's past toward the West and away from that of the eastern Islamic world, see Kaufman (2004).

2. Here I draw on Tahan's (2005) discussion of the museum during wartime along with artist Lamia Joreige's (2013) narrative that accompanied her installation documenting the history of the National Museum.

3. In everyday parlance, I found that people did use the term *civil war* (*al-harb al-ahliya*), although they assumed knowledge and understanding, on the part of those they were speaking with, of the regional dimensions of the war. In short, while the term *civil war* was regularly used, it was shorthand for a war of strategic global geopolitical significance that was fought by both domestic and regional players.

4. Article 9 of the constitution, for example, put rights and rulings on personal status (marriage, divorce, custody, adoption, inheritance) under the domain of religious sectarian communities (Traboulsi 2007, 90).

5. This idea of Lebanon's Muslim and Arab character represented an ideological compromise between the Muslim and Christian groups. The compromise was that Muslims would recognize the distinct nature of their country, marked by its sectarian political system and historic ties with the West, and that Christians would renounce any protective links to the European powers and affirm the Arab character of the country and its membership in the Arab world.

6. For more about the 1932 census and its enduring political significance, see Maktabi (1999).

7. See Jim Lehrer's PBS *News Hour* interview with President George W. Bush about the state of the war in Iraq (Bush, 2007).

8. The conflict also marked a new phase of U.S.-Lebanon relations and the progressively deeper level of U.S. engagement in both regional and Lebanese politics. See Gendzier (1997) for a careful reading of the 1958 conflict and its significance for U.S.-Lebanon relations.

9. See Traboulsi (2007, 160) for statistics on the cost of living between 1967 and 1975; Picard (1996, 94) on the unemployment rate in 1975; and Joseph (1983, 11) on the percentage of the population controlling the GNP. Khalaf (2002) provides a detailed picture of the economic situation in the prewar moment.

10. Indeed, the militias became large business enterprises during Lebanon's war. Among other activities, Traboulsi describes how the militias monopolized foreign trade, controlled the lottery business, participated in the narcotics traffic, developed a black market of imports to Syria, and engaged in "exchange services" with the bourgeoisie (protection money in return for import and export quotas or sheer profiteering). They became an integral part of that class, entering into close business partnerships with many of its members, especially in the flour and fuel trade (2007, 236–237).

11. These figures are from Peteet's (2005) study of Palestinians in Lebanon.

12. There are now twelve camps in Lebanon. See the UNRWA website (http://www.unrwa.org/) for information on the present situation.

13. Several anthropologists have documented the experience of Palestinians in Lebanon. See, for example, Allan (2013); Khalili (2007); and Peteet (2005).

14. Fisk's on-the-ground narrative *Pity the Nation* (2002) provides this kind of account by detailing the many stages, theaters, and episodes of the war. Haugbolle (2011) offers a thorough review of the historiography and memory of the war.

15. Haugbolle (2012) analyzes discourses about masculinity and militiamen during the war from various angles, including that of the "little militia man," who takes up arms in the midst of the transition from boyhood to manhood.

16. Here, I borrow Picard's term (1996, 149) to describe the divided political and administrative structure of the city during the war.

17. Israeli troops withdrew from South Lebanon in 2000, and a United Nations Interim Force in Lebanon zone was established in an area between the city of Tyre and the Israeli border. For more on Hizbullah and the Israeli withdrawal, see Norton (2000). Many Lebanese and Hizbullah dispute the Israeli occupation of a piece of land at the Lebanon-Syria border called Shebaa Farms, which Israel seized as part of the Golan Heights in the 1967 Arab-Israeli war. The possession of weapons by Hizbullah is an ongoing issue of political contention in Lebanon.

18. These figures are taken from Labaki and Abou Rjeily (1994, cited in Haugbolle 2011). In his review of the historiography and memory of the war across scholarly, literary, and artistic realms, Haugbolle (2011) discusses politicization surrounding the quantification of the war's casualties.

19. I have heard the ending of the war referred to as the moment when "people tired of fighting one another" but also as the moment when the powers-that-be tired of having the militias fight for them because the war had ceased to be profitable.

20. Although the Ta'if Accord increased the number of parliamentary seats from 99 to 108, in 1992 this number was changed to 128. Farid el Khazen discusses the rumors and lack of transparency surrounding the decision to establish 128 as the number of seats. The before Ta'if/after Ta'if parliamentary seat distribution looked like this: Maronite before Ta'if 30/after Ta'if 34, Greek Orthodox 11/14, Greek Catholic 6/8, Armenian Orthodox 4/5, Armenian Catholic 1/1, Protestant 1/1, Other Christians 1/1, total Christians 54/64; Sunni Muslims 20/27, Shi'i Muslims 19/27, Druze Muslims 6/8, Alawite Muslims 0/2, total Muslims 45/64. Total seats 99/128 (el Khazen 1994).

21. Amnesty International states: "No criminal investigations or prosecutions were initiated into mass human rights abuses that were committed with impunity during and after the 1975–1990 war. Abuses included killings of civilians; abductions and enforced disappearances of Palestinians, Lebanese and foreign nationals; and arbitrary detentions by various armed militias and Syrian and Israeli government forces." (*Lebanon—Amnesty International Report 2008*, accessed December 18, 2014, http://www.amnesty.org/en/region/lebanon/report-2008.)

22. For example, after returning from exile in France in 2005, Michel Aoun, a former Lebanese Army commander during Lebanon's war, became the leader of the Free

Patriotic Movement, a powerful political party with a substantial number of parliamentary seats. Nabih Berri, the current speaker of Parliament, led the Amal Movement during the war and the War of the Camps, in which several Palestinian refugee camps were besieged.

23. According to Mawad (2009), because the state gives schools the freedom to choose their own history textbooks, schools usually select one that is in line with their political and religious affiliation. This point highlights the ways in which political groups, some of which are framed through religion, are not only active in providing interpretations of present-day issues but serve as interpreters of the past.

24. This is the outlook that Peleikis (2006) encountered in her research with residents of a multisectarian village in the Shouf district of Mount Lebanon.

25. A significant amount of scholarly and artistic work also engages memories of the war. See, for example, *West Beirut* (1998); Larkin (2012); Makdisi (1999); Raad (2007). There are also ways in which the war is mined in popular culture for its irony-laced "cool" factor. For instance, in 2004, a bar named "1975" with waiters in camouflage, floor seating lined with sandbags, and graffiti-marked walls opened in the upscale nightlife district of Monot, a playground for the young and affluent.

26. The phrase in Arabic is *'amaru al-hajr bas ma 'amaru an-nas*.

27. The Syrian army first entered Lebanon in 1976 at the request of Lebanese President Suleiman Frangieh in the second year of the war, and the Arab League then gave Syria a mandate to retain troops in Lebanon with the objective of restoring peace. Syria became one of the major players in the war and asserted throughout that its actions were necessary to ensure the protection of both Syria and Lebanon.

28. Larkin describes one such plaque commemorating a martyr of the Syrian Social Nationalist Party in the Hamra neighborhood (2012, 104), and on walks through Ashrafieh and adjacent neighborhoods I have seen graffiti and banners honoring Bashir Gemayel, commander of the Lebanese Forces militia and senior member of the Phalange party, who was killed in 1982. For a discussion of martyrdom in the Lebanese context, see Volk (2010, 32–35).

29. In Larkin's research (2012) with members of what he calls the "post-memory" generation, the war was always present in their sentiments about political, civic, and social life.

30. Saree Makdisi argues, however, that the built fabric of downtown was damaged more by the reconstruction efforts than by the war itself: "It is estimated that, as the result of demolition, by the time reconstruction efforts began in earnest in 1993, approximately 80 percent of the structures in the downtown area had been damaged beyond repair, whereas only around a third held that status as a result of damage inflicted during the war itself" (1997, 674).

31. Prior to the war, downtown was a transportation hub with buses and service taxis lined up around Martyr's Square beside signs indicating their destinations. As an eclectic retail center, with sex-based commercial activity along with upmarket shops and restaurants, and as a point of origin for travel to other parts of Beirut, downtown was used by most of the city's residents. For more on downtown in the era prior to the long war

in Lebanon, see Khalaf (2006). For more on the early plans to rebuild downtown, see Beyhum (1992).

32. See Makdisi (1997, 667) for more on the demolition that was undertaken.

33. Residents and owners of property in the city center were offered modest compensation packages. Many of the displaced residents had fled to Beirut from rural areas during the war. Some residents were accused by Solidere of being "occupiers" or wartime militia members (or both) who filed illegitimate claims in order to receive compensation. According to the government, anyone who could present persuasive arguments or evidence showing that they were uprooted or relocated at least once during the war qualified for compensation from the Ministry of the Displaced and the Central Fund for the Displaced (Sawalha 2003, 276–278). The stated differences between the categories of "the displaced" and "the occupiers" were a subject of contention. In exchange for expropriation, property owners were offered Solidere shares. And it was stipulated that owners (or other interested parties) who wished to save buildings from demolition and instead recover and restore them would have to pay the company a 12 percent surcharge on the estimated value of the lot and be prepared to repair the building within two years.

34. Makdisi (1997) discusses the ancient archeological dimensions of the area.

35. For scholarly work on the downtown reconstruction and the politics of memory, see, for example, Haugbolle (2011); Makdisi (1997); Nagel (2002); Sawalha (2010); Schmid (2002); Yahya (2007).

36. For more on the notion of urbicide, see Fregonese (2009) on Beirut and Coward (2004) on Bosnia.

37. See Grodach (2002) on Bosnia and Lee (2013) and Nagle (2009) on Belfast.

38. See Hazbun (2008) for more on the effects of 9/11 on Arab tourism flows.

39. Volk (2010) provides a careful reading of the Hariri burial site and its visual iconography.

40. See Human Rights Watch, "Israel/Lebanon: Israeli Indiscriminate Attacks Kill Most Civilians," accessed December 18, 2014, http://www.hrw.org/news/2007/09/05/israellebanon-israeli-indiscriminate-attacks-killed-most-civilians.

41. For more about the threat of war debris, see the Lebanon section on the United Nations Office for Project Services website (http://www.unops.org).

42. Hermez (2012) conceptualizes this state of anticipating future war, which is shaped by past experiences of conflict in Lebanon.

CHAPTER 3: POLITICS AND PUBLIC SPACE

Parts of this chapter appeared in Kristin Monroe, "Youth Expressions of Class and Mobility," in *Everyday Life in the Muslim Middle East*, 3rd ed., ed. Donna Lee Bowen, Evelyn A. Early, and Becky Schulthies (Bloomington: Indiana University Press, 2014), 39–48. I thank Indiana University Press for permission to use this material.

1. In considering how everyday uses of the city can shed light on the workings of political sectarianism, I draw on de Certeau's notion of "spatial stories," which "have the function of founding and articulating spaces" (1984, 122).

2. For more on other divided cities, see Calame and Charlesworth (2009) on Belfast, Jerusalem, Mostar, and Nicosia, and see Allegra, Casaglia, and Rokem (2012) on the concept of the divided city.

3. Hizbullah expanded its role in government in 2005; for more, see Deeb and Harb (2013, 41).

4. These are young people from middle- and upper-class backgrounds who lived in neighborhoods of municipal Beirut and the northern suburbs. They shared many of the same leisure activities of the youth of the southern suburbs that Deeb and Harb (2013) describe; however, the subjects of piety and morality were not central in our conversations or in their talk about where they liked to have fun in the city.

5. Genberg (2002) also describes how boundaries between neighborhoods are drawn through various kinds of visual and aural sectarian markers.

6. See chapter 2 for a historical overview of the sectarian composition of the Lebanese government. Electoral reform has been a contentious and key political issue since the end of the civil and regional war. For more, see Arda Arsenian Ekmekji's report "Confessionalism and Electoral Reform in Lebanon," accessed December 18, 2014, http:// www.aspeninstitute.org/publications/confessionalism-electoral-reform-lebanon.

7. On the allocation of welfare resources in Lebanon, see Cammett (2014).

8. The American University of Beirut is a selective and expensive English-language private university founded by U.S. Protestant missionaries in the late nineteenth century. The majority of its students are from relatively privileged backgrounds in comparison with the average Lebanese.

9. For more on the mosque's construction and its relation to Sunni leadership in Lebanon, see Vloeberghs (2012).

10. Khalaf tracks these and other developments of the downtown environment in the post-1990 period (2006, 165).

11. For more on the relations of patronage that buttress the Lebanese political system, see Hamzeh (2001).

12. See, for example, Wedeen (1999) on the practice of public visual display of political leaders in Syria as a means of symbolic domination.

13. The singular, *dahiya* (suburb), is used to refer to the group of majority Shi'i areas in southern Beirut, and the plural, *al-dawahi* (suburbs), is used to refer to other suburban areas. For more on the formation of and nomenclature surrounding Dahiya, see Deeb and Harb (2013, 46–49).

14. Deeb and Harb discuss the area's lack of singularity in regard to these characteristics (2013, 178).

15. Deeb (2006) and Deeb and Harb (2013) provide in-depth ethnographic investigations of piety and space in Dahiya.

16. Fawaz, Harb, and Gharbieh describe these manifestations of neighborhood security in their report on "Living Beirut's Security Zones" (2012, 181–182).

17. The Phalange is a political party supported mainly by Maronite Christians. It played a major role in Lebanon's war, and after a decline in the late 1980s and 1990s reemerged in the early 2000s. It is a member of the March 14th Alliance.

18. Fawaz, Harb, and Gharbieh discuss this as a form security intended to counter the threat posed by intercommunal riots, which generally occur at the intersection of different territories (2012, 181).
19. See chapter 4 for further discussion of security processes in public space.
20. Years after this conversation with Layla, on May 22, 2007, a bomb did explode late at night in the parking lot of ABC mall, killing one and injuring nine others. Less than twenty-four hours later, a bomb went off in Verdun, near the Dunes shopping center, injuring ten, including two children.
21. Kegels (2007) reports similar findings about privileged young Beirutis continuing to go out seeking nightlife during the 2005–2006 period of bombings.
22. This paradoxical image was expressed in the photo that won the World Press Photo Award in 2007. Widely discussed in the Western media, the photo depicts fashionable young people driving by and taking photos of destroyed buildings in the southern suburbs of Beirut.
23. For more on the state surveillance of young people in other contexts within the region, see Varzi's (2006) description of how the public conduct of Tehranian youth is surveiled and policed in a city "inhabited by the state" and Ghannam's (2013) discussion of male working-class youth in Cairo and their encounters with state security forces.
24. In addition to this sense of physical safety that arises from sectarian ties, other forms of security are cultivated through sectarian affiliation. As Sawalha found, the government's failure to provide city residents with necessary urban services—such as electricity, building and elevator safety, and tenants' rights—forces Beirutis to rely on sectarian connections to meet their needs (2010, 64).
25. For example, a French first name is often understood to identify an individual as Christian, while males named Ali or Hussein (prominent figures in the Shi'i hagiography) are thought to indicate a Shi'i family background
26. There is youth activism against sectarianism in Beirut, and this movement has been energized by the opening in the Mar Mikhael neighborhood of bookstores, galleries, and cafes that serve as a kind of launching pad and organizing base.
27. Interestingly, another report found that Lebanese youth showed "high levels of sectarian bias (in-group favoritism) along with low levels of acceptance of inter-sectarian relationships" (Harb 2010, 17–18).

CHAPTER 4: SECURING BEIRUT

Parts of this chapter appeared in Kristin V. Monroe, "Being Mobile in Beirut," *City & Society* 23, no. 1 (2011): 91–111. I thank Wiley for permission to use this material.
1. This intensification of security for the protection of prominent and public figures is but one of the five modalities of security in Beirut that Fawaz, Harb, and Gharbieh (2012) outline.
2. For more on the relationship between automobility and warfare, see Davis (2007).

3. On gated residences, surveillance, and private security in urban sites, see Caldeira (2000); Davis (1992); Low (2004); and Zhang (2010). On the role of vigilante groups performing extrastate security, see Goldstein (2012).
4. For more on these modes of collaborative policing between private and state-based forces, see Yarwood (2007) on multiagency policing and Jones and Newburn (2006) on plural policing.
5. Drawing on Allen Feldman's (2004) formulation of "securocratic warfare," I use the term *securocracy* to refer to the ways in which heightened anxieties about the security of high-profile individuals appear not only as a spatial phenomenon but also as part of broader political strategy.
6. Writings about SUVs have focused on their materiality (Miller 2007), how their size and weight create physical dangers for other users of the road (Jain 2002), and how they encourage an atomistic model of citizenship in which drivers inhabit buffer zones that limit their contact with others (Mitchell 2005).
7. Geographer Louise Amoore (2011) refers to this calculation as the "ontology of association."
8. These skills and tactics bring to mind the "ways of operating" de Certeau (1984) described as part of the everyday practice of making one's way across spatial terrain.
9. See Fawaz, Gharbieh, and Harb (2009) for a vivid picture of the complexities surrounding Beirut's security context.
10. U.S. Homeland Security Presidential Directive 3 was issued in March 2002 and requires the attorney general in consultation with the assistant to the president for Homeland Security to assign the United States one of five "threat conditions" identified by a description and corresponding color. See "Homeland Security Presidential Directive-3," George W. Bush White House Archives, accessed December 19, 2014, http://georgewbush-whitehouse.archives.gov/news/releases/2002/03/20020312-5.html.
11. Moreover, the increased militarization of the city not only connected this period with that of the civil war but also masculinized public space. Urban space was increasingly governed by men with guns, and moving through the city entailed regular interactions with them. These interactions were gendered, for example, through the use of aggressive verbal and physical action and the assertion of a masculinist protectionism particularly in relation to (some) women and children.

CHAPTER 5: THE CHAOS OF DRIVING

Parts of this chapter appeared in Kristin V. Monroe, "Being Mobile in Beirut," *City & Society* 23, no. 1 (2011): 91–111. I thank Wiley for permission to use this material.
1. Lebanese Arabic is peppered with French words. Some Lebanese, for instance, use the French word *accident* rather than the Arabic *hadith*.
2. Statistics I obtained from the traffic-police division of the Internal Security Forces (ISF) recorded 6,508 injuries and 513 fatalities resulting from traffic accidents in 2009. In interpreting such statistics, the Youth Association for Social Awareness (YASA)

highlights the size and population figures of Lebanon, the existence of underreporting, and the fact that the traffic police investigate neither damages-only accidents nor those involving army and police vehicles, which may constitute about 15 percent of all road-related fatalities according to the Scientific Research Foundation, YASA website, accessed December 10, 2014, http://www.yasa.org/en/Sectiondet.aspx?id=10&id2=371.

3. Buses, like service taxis, are also flagged down as there are no fixed stops along the bus routes. The routes themselves can vary as drivers often choose to avoid traffic or construction by taking a different "series of streets that are not along their typical route. Buses came to replace the tram system in the early 1960s.

4. I use the male pronoun for service taxi drivers because in my time living in Beirut, I never saw a female taxi driver, although a friend insists that there are a few. She told me that she had seen two after living in Beirut for thirteen years.

5. Statistic from the World Bank's country profile of Lebanon, accessed October 12, 2010, http://web.worldbank.org/. The most current annual per capita income data from the World Bank is US$17,090.

6. Examples of cultural productions highlighting the intergroup sociality that takes place in service taxis are the feature film *Taxi Ballad* (2012), the documentary film *Taxi Talk* (2009), and a collection of essays in Arabic, *Beirut bil Service* (Krideyah 2009).

7. Stoller (1982) describes the bush taxi as a rich ethnographic site through which the anthropologist learns to interpret the signs that constitute the discourse of social action.

8. While perusing issues of *L'Orient-Le Jour* from the 1960s, I found articles about the service taxi system that confirmed comments by older Beirutis that "before the [civil and regional] war, the taxis were more orderly" because they followed fixed routes to different sectors of the city.

9. Most upper-income residents I met with said they rarely or never took service taxis, and women from this same class background voiced concerns about the safety of these taxis with reference to both themselves and their preteen and teenaged daughters.

10. Notar (2012b) discusses this phenomenon in the context of urban China.

11. As I describe here in the context of Beirutis' various understandings of the chaos of driving, Anderson (1983) saw this "deep, horizontal comradeship" as being produced in spite of the actual inequality and exploitation that may prevail in a national community.

12. Curiously, a French Mandate-era Lebanese newspaper cartoon I came across encouraged me to think about the origins of this discourse about "becoming" or "being" Lebanese through the way in which one drove. The 1922 cartoon from *Al-Ma'aarad*, entitled "How to Transgress Rules," depicted a curbside lined with parked cars with a sign declaring "Parking of cars and carriages prohibited."

13. Although emerging from an ecological perspective, the topic of social disorganization also engaged early urban studies scholars of the Chicago School. See, for example, Park, Burgess, and McKenzie's now-classic *The City* (1925).

14. *As-Safir*, April 10, 1995, 5.

15. From "Fadlallah lors de sa Réception d'une Délégation de l'organisation YASA: Livrer un Permis de Conduire à un Non Méritant est Illicite et est Une Fraude!"

accessed June 14, 2010. http://www.yasa.org/fr/Section.aspx?id=6 (article now deleted from site).
16. From "Fadlallah a Reçu le Député Pierre Dakkash et le Rassemblement des Jeunes pour la Sensibilisation Sociale," accessed June 14, 2010. http://www.yasa.org/fr/Section.aspx?id=6 (article now deleted from site).
17. Here I invoke Lefebvre's idea that cities have collective rhythms that are determined by the varied and contradictory forms of alliances that human groups create and in which class and political relations intervene ([1968] 1996, 234).

CHAPTER 6: "THERE IS NO STATE"

1. The ISF is the domestic police force under the authority of the Ministry of the Interior; among other duties, it administers traffic policing. Although separate from the armed forces, the ISF is paramilitary in its organization, decorum, and uniforms.
2. Baroud increased traffic fines, the number of traffic police, and the allocation of funds for traffic management, for example, while expanding the use of radar equipment. See Sikimic (2010).
3. Ethnographic attention is now being paid to the everyday bureaucratic activities and actors that make up the state. See, for example, Gupta (2012) and Navaro-Yashin (2012).
4. For examples of this kind of weak-state analysis of Lebanon, see Hanf (1993); Pan (2006); Rotberg (2003).
5. In the Lebanese context, however, Fregonese (2012) argues that a conception of a single sovereignty is ill-fitting as Lebanon is characterized by multiple sovereignties constituted by both state and nonstate actors.
6. The 16th Brigade was a special police force "of the most modern means" created in 1959 during Fouad Chehab's presidency to deal with emergencies related to internal security (Malsagne 2008).
7. In saying "there is no state," the policeman recalled to me the cynicism of the state administration workers in the Turkish Republic of Northern Cyprus—a self-declared but internationally unrecognized state—which Navaro-Yashin (2012) refers to as being a "made-up state." In his verbiage, the police officer may also have sought to distance himself, as the Indian government workers in Gupta and Sharma's research did (2006, 286), from the well-circulated image of the lazy, inefficient, and corrupt state worker and thereby to remove himself, as it were, from the sullied reputation of the state and its functionaries like the traffic police.
8. Police training takes place in military schools where recruits participate in a month-long specialization program like policing traffic. For more on Lebanon's security apparatus, see Nashabe (2009).
9. Although officially the ISF is not an exclusively male organization and while public relations materials for the ISF include images of women and I saw several when visiting ISF offices, on the streets I have only rarely seen female members of the ISF.
10. USAID provided this assistance. See, for example, usaid.gov/pdf_docs/PDACR061.pdf, accessed May 16, 2012.

11. These numbers ebb and flow in response to domestic and regional political tensions and dynamics. The highest tourism levels thus far in the post-civil and regional war era—two million visitors—were recorded in 2009 (Kourchid 2009).
12. Here I make reference to Gupta and Ferguson's (2002) analysis of the spatiality of the state and specifically its vertical and encompassing dimensions.
13. As noted, in 2004, there were only a few traffic lights downtown. By summer 2010, additional traffic lights had appeared and more drivers obeyed them. However, at the most unwieldy intersections policemen still controlled traffic.
14. For analyses of networks of favoritism and patronage in Lebanon and beyond, see Cunningham and Sarayrah (1994); Joseph (1983); Lomnitz (1971); and Makhoul and Harrison (2004).
15. On wasta's effect on the business climate, see Loewe, Blume, and Speer (2008); on possibilities for career advancement, see Tlaiss and Kauser (2011); and on nations' economic competitiveness, see Mohamed and Hamdy (2008).
16. In addition to buying licenses, many Lebanese criticize the ease with which people obtain them. Journalist Habib Battah's June 18, 2013, blogspot referred to Lebanon's Department of Motor Vehicles as "The Most Dangerous Place in Lebanon"—a place where "thousands of killers are licensed each year" after taking a sham of a driving test. "The Most Dangerous Place in Lebanon," accessed December 19, 2014, http://www.beirutreport.com/2013/06/the-most-dangerous-place-in-lebanon.html.
17. This is due to the potential for a male-male encounter in this cultural context of policing, in which the hegemonic view of women is as nonthreatening, to become a violent power play.
18. On luxury license plates, see Braun (2004).

CONCLUSION

1. Nasrallah made this announcement on May 25, 2013. The day after his speech, rocket fire attacked Hizbullah-affiliated areas in Beirut and Hermel in eastern Lebanon. Syrian rebels were blamed for what was widely held to be a retaliatory attack. For more on the speech and the attacks, see "Nasrallah on Syria: The Battle Is Ours," Al-Monitor website, accessed December 19, 2014, http://www.al-monitor.com/pulse/originals/2013/05/nasrallah-hezbollah-syria-speech-rockets.html#.
2. This statistic is from the U.N. refugee agency, UNHCR, "Syria Regional Refugee Response," accessed December 16, 2014, http://data.unhcr.org/syrianrefugees/country.php?id=122.
3. For more on how the refugee crisis has affected Lebanon's economic and labor situation, see the International Labour Organization report (2014, 36).
4. Here, I borrow Neil Smith's phrasing, from his entreaty for increased attention to class in contemporary social theory and geography, about how "class is thickly written through cultural, political, and economic landscapes" (2000, 1012).

5. And, yet, members of government do make these kinds of claims, as Interior Minister Marwan Charbel did in summer 2013, when, amid the rising political violence associated with the Syrian civil war, he assured Lebanese that the state protects all Lebanese citizens (see, for example, "The State Embraces All Parties," National News Agency website, accessed December 19, 2014, http://www.nna-leb.gov.lb/en/show-news/14298/) and rejects "private security" checkpoints and patrols established by groups other than state security and military units (see, for example, "The State Rejects Private Security Measures," *Daily Star*, accessed December 19, 2014, http://www.dailystar.com.lb/News/Lebanon-News/2013/Sep-12/230967-charbel-state-rejects-private-security-measures.ashx#axzz2kGXw2ibv).

6. Abrams proposes alternative modes of conceptualizing the state in his landmark essay "Notes on the Difficulty of Studying the State" ([1977] 1988, 82).

7. Here I reference Timothy Mitchell's critique (1991) of theorization of the state as a coherent, autonomous entity that operates in a separate realm from that of society.

8. Lefebvre wrote about the character—and social promise—of the city in several essays and addressed the nature of the urban fabric as a place of encounter in "The Right to the City" ([1968] 1996, 158).

REFERENCES

Abdou-Hodeib, Toufoul. 2011. "Taste and Class in Late Ottoman Beirut." *International Journal of Middle East Studies* 43 (3): 475–492.

Abrams, Philip. (1977) 1988. "Notes on the Difficulty of Studying the State." *Historical Sociology* 1 (1): 58–89.

Abu-Lughod, Lila. 1990. "The Romance of Resistance: Tracing Transformations of Power through Bedouin Women." *American Ethnologist* 17 (1): 41–55.

Adnan, Etel. 1982. *Sitt Marie Rose*. Translated by Georgina Kleege. Sausalito, CA: Post-Apollo Press.

Allan, Diana. 2013. *Refugees of the Revolution: Experiences of Palestinian Exile*. Stanford, CA: Stanford University Press.

Allegra, Marco, Anna Casaglia, and Jonathan Rokem. 2012. "The Political Geographies of Urban Polarization: A Critical Review of Research on Divided Cities." *Geography Compass* 6 (9): 560–574.

Allen, Lori. 2008. "Getting by the Occupation: How Violence Became Normal during the Second Palestinian Intifada." *Cultural Anthropology* 23 (3): 453–487.

Amar, Paul. 2013. *The Security Archipelago: Human-Security States, Sexuality Politics, and the End of Neoliberalism*. Durham, NC: Duke University Press.

Amoore, Louise. 2011. "Data Derivatives: On the Emergence of a Security Risk Calculus for Our Times." *Theory, Culture & Society* 28 (6): 24–43.

Anderson, Ben. 2010. "Preemption, Precaution, Preparedness: Anticipatory Action and Future Geographies." *Progress in Human Geography* 34 (6): 777–798.

Anderson, Benedict. 1983. *Imagined Communities: Reflections on the Origin and Spread of Nationalism*. London: Verso.

Aoun, Alisar, Maya Abou-Zeid, Isam Kaysi, and Cynthia Myntti. 2013. "Reducing Parking Demand and Traffic Congestion at the American University of Beirut." *Transport Policy* 25: 52–60.

Aretxaga, Begoña. 2003. "Maddening States." *Annual Review of Anthropology* 32: 393–410.

The Autostrad: A Mezé Culture—Lebanon and Auto-mobility. 2003. Louaize, Lebanon: Faculty of Architecture, Art, and Design, Notre Dame University.

Barakat, Halim. 1973. "Social and Political Integration in Lebanon: A Case of Social Mosaic." *Middle East Journal* 27 (3): 301–318.

"Beirut Tourist Police Attempt to Silence Automobile Horns." 1964. *Times of London*, December 6.

Bell, Colleen. 2006. "Surveillance Strategies and Populations at Risk: Biopolitical Governance in Canada's National Security Policy." *Security Dialogue* 37 (2): 147–165.

Belloncle, Edouard. 2006. "Prospects of Security Sector Reform in Lebanon." *Journal of Security Sector Management* 4 (4). http://www.ssronline.org/jofssm/index.cfm.

Beyhum, Nabil. 1992. "Beirut's Three Reconstruction Plans." *Beirut Review*, no. 4 (Fall): 43–57.
Bourdieu, Pierre. 1977. *Outline of a Theory of Practice*. Translated by Richard Nice. Cambridge: Cambridge University Press.
Braun, Amy. 2004. "Fewer Digits on License Plate, More Kudos for Your Car." *Daily Star*, April 15. http://www.dailystar.com.lb/News/Lebanon-News/2004/Apr-15/2019-fewer-digits-on-license-plate-more-kudos-for-your-car.ashx.
Bush, George W. 2007. Interview with Jim Lehrer. *PBS News Hour*, January 16.
Calame, Jon, and Esther Charlesworth. 2009. *Divided Cities: Belfast, Beirut, Jerusalem, Mostar, and Nicosia*. Philadelphia: University of Pennsylvania Press.
Caldeira, Teresa. 2000. *City of Walls: Crime, Segregation, and Citizenship in São Paulo*. Berkeley: University of California Press.
———. 2012. "Imprinting and Moving Around: New Visibilities and Configurations of Public Space in São Paulo." *Public Culture* 24 (2): 385–419.
Cammett, Melani. 2014. *Compassionate Communalism: Welfare and Sectarianism in Lebanon*. Ithaca, NY: Cornell University Press.
Chabban, F. B., G. M. Aoub, and M. Oulabi. 2001. "Preliminary Evaluation of Selected Transport-Related Pollutants in the Ambient Atmosphere of the City of Beirut, Lebanon." *Water, Air, and Soil Pollution* 126 (1/2): 53–62.
Chatterji, Roma, and Deepak Mehta. 2007. *Living with Violence: An Anthropology of Events and Everyday Life*. London: Routledge.
Comaroff, Jean, and John Comaroff. 2007. "Law and Disorder in the Postcolony." *Social Anthropology* 15 (2): 133–152.
Coward, Martin. 2004. "Urbicide in Bosnia." In *Cities, War, and Terrorism: Towards an Urban Geopolitics*, edited by Stephen Graham, 154–171. Malden, MA: Blackwell.
Cunningham, Robert B., and Yasin K. Sarayrah. 1994. "Taming Wasta to Achieve Development." *Arab Studies Quarterly* 16 (3): 29–41.
Czeglédy, André. 2004. "Getting Around Town: Transportation and the Built Environment in Post-apartheid South Africa." *City & Society* 16 (2): 63–92.
DaMatta, Roberto. 1991. "Do You Know Who You're Talking To? The Distinction between Individual and Person in Brazil." In *Carnivals, Rogues, and Heroes: An Interpretation of the Brazilian Dilemma*, translated by John Drury, 137–197. Notre Dame, IN: University of Notre Dame Press.
Davie, May. 2003. "Beirut and the 'Etoile' Area: An Exclusively French Project?" In *Urbanism: Imported or Exported? Native Aspirations and Foreign Plans*, edited by Joe Nasr and Mercedes Volait, 171–230. Chichester, UK: Wiley-Academy.
Davis, Mike. 1992. *City of Quartz: Excavating the Future in Los Angeles*. New York: Vintage Books.
———. 2007. *Buda's Wagon: A Brief History of the Car Bomb*. New York: Verso.
de Certeau, Michel. 1984. *The Practice of Everyday Life*. Translated by Steven F. Rendall. Berkeley: University of California Press.

Deeb, Lara. 2006. *An Enchanted Modern: Gender and Public Piety in Shi'i Lebanon*. Princeton, NJ: Princeton University Press.

Deeb, Lara, and Mona Harb. 2013. *Leisurely Islam: Negotiating Geography and Morality in Shi'ite South Beirut*. Princeton, NJ: Princeton University Press.

de Koning, Anouk. 2009. *Global Dreams: Class, Gender, and Public Space in Cosmopolitan Cairo*. Cairo: American University in Cairo Press.

Dueck, Jennifer. 2010. *The Claims of Culture at Empire's End: Syria and Lebanon under French Rule*. Oxford: Oxford University Press.

———. 2012. "Flourishing in Exile: French Missionaries in Syria and Lebanon under Mandate Rule." In *In God's Empire: French Missionaries and the Modern World*, edited by Owen White and J. P. Daughton, 151–172. New York: Oxford University Press.

el Khazen, Farid. 1994. "Lebanon's First Postwar Parliamentary Elections, 1993." *Middle East Policy* 3 (1): 120–136.

Elsheshtawy, Yasser, ed. 2008. *The Evolving Arab City: Tradition, Modernity and Urban Development*. Abingdon, UK: Routledge.

Fawaz, Leila. 1983. *Merchants and Migrants in Nineteenth-Century Beirut*. Cambridge, MA: Harvard University Press.

Fawaz, Mona. 2009a. "Hezbollah as Urban Planner? Questions to and from Planning Theory." *Planning Theory* 8 (4): 323–334.

———. 2009b. "Neoliberal Urbanity and the Right to the City: A View from Beirut's Periphery." *Development and Change* 40 (5): 827–852.

———. 2009c. "The State and the Production of Illegal Housing: Public Practices in Hayy el Sellom, Beirut-Lebanon." In *Comparing Cities: The Middle East and South Asia*, edited by Kamran Asdar Ali and Marina Rieker, 197–220. Oxford: Oxford University Press.

Fawaz, Mona, Ahmad Gharbieh, and Mona Harb. 2009. *Beirut: Mapping Security*. Diwan Series, International Architecture Biennale Rotterdam. http://www.academia.edu/9224504/BEIRUT_MAPPING_SECURITY_.

Fawaz, Mona, Mona Harb, and Ahmad Gharbieh. 2012. "Living Beirut's Security Zones: An Investigation of the Modalities and Practice of Urban Security." *City & Society* 24 (2): 173–195.

Feldman, Allen. 2004. "Securocratic Wars of Public Safety: Globalized Policing as Scopic Regime." *Interventions: International Journal of Postcolonial Studies* 6 (3): 330–350.

Fenster, Tovi, and Hanaa Hamdan-Saliba. 2013. "Gender and Feminist Geographies in the Middle East." *Gender, Place & Culture: A Journal of Feminist Geography* 20 (4): 528–546.

Ferguson, James. 2005. "Seeing Like an Oil Company: Space, Security, and Global Capital in Neoliberal Africa." *American Anthropologist* 107 (3): 377–382.

Fisk, Robert. 2002. *Pity the Nation: The Abduction of Lebanon*. New York: Nation Books.

Foucault, Michel. 2007. *Security, Territory, Population: Lectures at the Collège de France 1977–1978*. Translated by Graham Burchell. New York: Picador.

Fregonese, Sara. 2009. "The Urbicide of Beirut? Geopolitics and the Built Environment in the Lebanese Civil War (1975–1976)." *Political Geography* 28 (5): 309–318.

———. 2012. "Beyond the 'Weak State': Hybrid Sovereignties in Beirut." *Environment and Planning D: Society and Space* 30 (4): 655–674.

Fricke, Adrienne. 2005. "Forever Nearing the Finish Line: Heritage Policy and the Problem of Memory in Postwar Beirut." *International Journal of Cultural Property* 12 (2): 163–181.

Gates, Carolyn F. 1998. *The Merchant Republic of Lebanon: Rise of an Open Economy*. Oxford: Centre for Lebanese Studies.

Genberg, Daniel. 2002. "Borders and Boundaries in Post-war Beirut." In *Urban Ethnic Encounters: The Spatial Consequences*, edited by Freek Colombijn and Aygen Erdentug, 81–96. London: Routledge.

Gendzier, Irene L. 1997. *Notes from the Minefield: United States Intervention in Lebanon and the Middle East 1945–1958*. New York: Columbia University Press.

Ghannam, Farha. 2002. *Remaking the Modern: Space, Relocation, and the Politics of Identity in a Global Cairo*. Berkeley: University of California Press.

———. 2013. *Live and Die Like a Man: Gender Dynamics in Urban Egypt*. Stanford, CA: Stanford University Press.

Giddens, Anthony. 1984. *The Constitution of Society: Outline of the Theory of Structuration*. Berkeley: University of California Press.

Glasze, Georg, and Abadallah Alkhayyal. 2002. "Gated Housing Estates in the Arab World: Case Studies in Lebanon and Riyadh, Saudi Arabia." *Environment and Planning B: Planning and Design* 29 (3): 321–336.

Goldstein, Daniel M. 2010. "Toward a Critical Anthropology of Security." *Current Anthropology* 51 (4): 487–517.

———. 2012. *Between Security and Rights in a Bolivian City*. Durham, NC: Duke University Press.

Graham, Stephen. 2010. *Cities under Siege: The New Military Urbanism*. London: Verso.

Grodach, Carl. 2002. "Reconstituting Identity and History in Post-war Mostar, Bosnia-Herzegovina." *City* 6 (1): 61–82.

Gupta, Akhil. 2012. *Red Tape: Bureaucracy, Structural Violence, and Poverty in India*. Durham, NC: Duke University Press.

Gupta, Akhil, and James Ferguson. 2002. "Spatializing States: Toward an Ethnography of Neoliberal Governmentality." *American Ethnologist* 29 (4): 981–1002.

Gupta, Akhil, and Aradhana Sharma. 2006. "Globalization and Postcolonial States." *Current Anthropology* 47 (2): 277–307.

Gusterson, Hugh, and Catherine Besteman, eds. 2010. *The Insecure American: How We Got Here and What We Should Do about It*. Berkeley: University of California Press.

Hall, Stuart. 1985. "Signification, Representation, Ideology: Althusser and the Post-structuralist Debates." *Critical Studies in Mass Communication* 2: 91–114.

Hammami, Rema. 2004. "On the Importance of Thugs: The Moral Economy of a Checkpoint." *Jerusalem Quarterly* 22 & 23: 16–28.

Hamzeh, A. Nizar. 2001. "Clientalism, Lebanon: Roots and Trends." *Middle Eastern Studies* 37 (3): 167–178.

Hanafi, Sari, Jad Chaaban, and Karin Seyfert. 2012. "Social Exclusion of Palestinian Refugees in Lebanon: Reflections on the Mechanisms That Cement Their Persistent Poverty." *Refugee Survey Quarterly* 31 (1): 34–53.

Hanf, Theodor. 1993. *Coexistence in Wartime Lebanon: Decline of a State and Rise of a Nation.* London: Tauris.

Hannam, Kevin, Mimi Sheller, and John Urry. 2006. "Mobilities, Immobilities, and Moorings." *Mobilities* 1 (1): 1–22.

Hannerz, Ulf. 1980. *Exploring the City: Inquiries toward an Urban Anthropology.* New York: Columbia University Press.

Hanssen, Jens. 2005. *Fin de Siècle Beirut: The Making of an Ottoman Provincial Capital.* Oxford: Oxford University Press.

Harb, Charles. 2010. "Describing the Lebanese Youth: A National and Psycho-social Survey." Youth in the Arab World Working Paper Series, no. 3. Beirut: Issam Fares Institute for Public Policy and International Affairs, American University of Beirut.

Harb, Mona. 2007. "Deconstructing Hizballah and Its Suburb." *Middle East Report* 242 (Spring): 12–17.

Harvey, David. 2003. "The Right to the City." *International Journal of Urban and Regional Research* 27 (4): 939–941.

Haugbolle, Sune. 2010. *War and Memory in Lebanon.* Cambridge: Cambridge University Press.

———. 2011. "The Historiography and Memory of the Lebanese Civil War 1975–1990." *Online Encyclopedia of Mass Violence.* http://www.massviolence.org/The-historiography-and-the-memory-of-the-Lebanese-civil-war.

———. 2012. "The (Little) Militia Man: Memory and Militarized Masculinity in Lebanon." *Journal of Middle East Women's Studies* 8 (1): 115–139.

Hayden, Dolores. 1982. *The Grand Domestic Revolution: A History of Feminist Designs for American Homes, Neighborhoods, and Cities.* Cambridge, MA: MIT Press.

Hazbun, Waleed. 2008. *Beaches, Ruins, Resorts: The Politics of Tourism in the Arab World.* Minneapolis: University of Minnesota Press.

Hermez, Sami. 2012. "'The War Is Going to Ignite': On the Anticipation of Violence in Lebanon." *PoLAR: Political and Legal Anthropology Review* 35 (2): 327–344.

Holston, James, and Arjun Appadurai. 1996. "Cities and Citizenship." *Public Culture* 8 (2): 187–204.

Inhorn, Marcia C. 2012. *The New Arab Man: Emergent Masculinities, Technologies, and Islam in the Middle East.* Princeton, NJ: Princeton University Press.

International Labour Organization. 2014. "Assessment of the Impact of Syrian Refugees in Lebanon and Their Employment Profile." International Labour Organization Regional Office for the Arab States, http://www.ilo.org/wcmsp5/groups/public/@arabstates/@ro-beirut/documents/publication/wcms_240134.pdf, accessed 6/17/15.

Isin, Engin F. 2002. *Being Political: Genealogies of Citizenship*. Minneapolis: University of Minnesota Press.

Jain, S. Lochlann. 2002. "Urban Errands: The Means of Mobility." *Journal of Consumer Culture* 2 (3): 385–404.

Jegnathan, Pradeep. 2002. "Walking through Violence: 'Everyday Life and Anthropology.'" In *Everyday Life in South Asia*, edited by Diane P. Mines and Sarah Lamb, 357–365. Bloomington: Indiana University Press.

Jones, Trevor, and Tim Newburn, eds. 2006. *Plural Policing: A Comparative Perspective*. London: Routledge.

Joreige, Lamia. 2013. *Under-writing Beirut-Mathaf*. Edited by Nadia Al Issa. Translated by Jacques Aswad. Multimedia installation and publication commissioned by Sharjah Art Foundation. http://www.lamiajoreige.com/publications.php.

Joseph, Suad. 1983. "Working Class Women's Networks in a Sectarian State: A Political Paradox." *American Ethnologist* 10 (1): 1–22.

Kanna, Ahmed. 2011. *Dubai: The City as Corporation*. Minneapolis: University of Minnesota Press.

Katz, Jack. 1999. *How Emotions Work*. Chicago: University of Chicago Press.

Kaufman, Asher. 2004. *Reviving Phoenicia: The Search for Identity in Lebanon*. London: Tauris.

Kegels, Nicolien. 2007. "Nothing Shines as a Beirut Night." *Etnofoor* 20 (2): 87–101.

Khalaf, Samir. 1985. "Social Structure and Urban Planning in Lebanon." In *Property, Social Structure, and Law in the Modern Middle East*, edited by Ann Elizabeth Mayer, 213–235. Albany: State University of New York Press.

———. 2002. *Civil and Uncivil Violence in Lebanon: A History of the Internationalization of Communal Conflict*. New York: Columbia University Press.

———. 2006. *Heart of Beirut: Reclaiming the Bourj*. London: Saqi Books.

Khalaf, Samir, and Per Kongstad. 1973. *Hamra of Beirut: A Case of Rapid Urbanization*. Leiden: Brill.

Khalili, Laleh. 2007. *Heroes and Martyrs of Palestine: The Politics of National Commemoration*. Cambridge: Cambridge University Press.

Khouri, Rami. 2011. "Youth Identities and Values: An Abundance of Identities in Evolving Societies." In *A Generation on the Move: Insights into the Conditions, Aspirations, and Activism of Arab Youth*. Beirut: UNICEF and Issam Fares Institute for Public Policy & International Affairs, American University of Beirut. http://apyouthnet.ilo.org/resources/a-generation-on-the-move-insights-into-the-conditions-aspirations-and-activism-of-arab-youth.

Kivland, Chelsey L. 2012. "Unmaking the State in 'Occupied' Haiti." *PoLAR: Political and Legal Anthropology Review* 35 (2): 248–270.

Korfali, Samira Ibrahim, and May Massoud. 2003. "Assessment of Community Noise Problem in Greater Beirut Area, Lebanon." *Environmental Monitoring and Assessment* 84 (3): 203–218.

Kourchid, Maya. 2009. "Summer Tourism." *NOW Lebanon*, July 7. http://www.nowlebanon.com/NewsArchiveDetails.aspx?ID=102653.

Krideyah, Aman. 2009. *Beirut bil Service*. Paris: Eastern Library.
Krijnen, Marieke, and Mona Fawaz. 2010. "Exception as the Rule: High-End Developments in Neoliberal Beirut." *Built Environment* 36 (2): 117–131.
Labaki, Boutros, and Khalil Abou Rjeily, eds. 1994. *Bilan des Guerres du Liban 1975–1990*. Paris: L'Harmattan.
Larkin, Craig. 2012. *Memory and Conflict in Lebanon: Remembering and Forgetting the Past*. Abingdon, UK: Routledge.
Lee, Adele. 2013. "Introduction: Post-conflict Belfast." *City* 17 (4): 523–525.
Lefebvre, Henri. [1968] 1996. *Writings on Cities*. Translated and edited by Eleonore Kofman and Elizabeth Lebas. Oxford: Blackwell.
Loewe, Markus, Jonas Blume, and Johanna Speer. 2008. "How Favoritism Affects the Business Climate: Empirical Evidence from Jordan." *Middle East Journal* 62 (2): 259–276.
Logan, John R., and Harvey L. Molotch. 1987. *Urban Fortunes: The Political Economy of Place*. Berkeley: University of California Press.
Lomnitz, Larissa. 1971. "Reciprocity of Favors in the Urban Middle Class of Chile." In *Studies in Economic Anthropology*, edited by George Dalton, 93–107. Washington, DC: American Anthropological Association.
Low, Setha M. 2004. *Behind the Gates: Life, Security, and the Pursuit of Happiness in Fortress America*. New York: Routledge.
Lutz, Catherine. 2014. "The U.S. Car Colossus and the Production of Inequality." *American Ethnologist* 41 (2): 232–245.
Maasri, Zeina. 2009. *Off the Wall: Political Posters of the Lebanese Civil War*. London: Tauris.
Makdisi, Jean. 1999. *Beirut Fragments: A War Memoir*. New York: Persea Books.
Makdisi, Saree. 1997. "Laying Claim to Beirut: Urban Narrative and Spatial Identity in the Age of Solidere." *Critical Inquiry* 23 (3): 660–705.
———. 2006. "Beirut, a City without History?" In *Memory and Violence in the Middle East and North Africa*, edited by Ussama Makdisi and Paul A. Silverstein, 201–215. Bloomington: Indiana University Press.
Makdisi, Ussama. 2000. *The Culture of Sectarianism: Community, History, and Violence in Nineteenth-Century Ottoman Lebanon*. Berkeley: University of California Press.
Makhoul, Jihad, and Lindsey Harrison. 2004. "Intercessory Wasta and Village Development in Lebanon." *Arab Studies Quarterly* 26 (3): 25–41.
Maktabi, Rania. 1999. "The Lebanese Census of 1932 Revisited: Who Are the Lebanese?" *British Journal of Middle Eastern Studies* 26 (2): 219–241.
Malsagne, Stéphan. 2008. "Foud Chehab (1902–1973): Contribution à l'étude d'une figure historique majeure du Liban contemporain." PhD diss., Université de Paris I Pantheon-Sorbonne.
Mansel, Philip. 2010. *Levant: Splendour and Catastrophe on the Mediterranean*. New Haven, CT: Yale University Press.
Massey, Doreen. 1994. *Space, Place, and Gender*. Minneapolis: University of Minnesota Press.

Mawad, Dalal. 2009. "Lebanon's History Awaits Its Textbook." *Daily Star*, November 20. http://www.dailystar.com.lb/Opinion/Commentary/2009/Nov-20/118255-lebanons-history-awaits-its-textbook.ashx.

Meinig, D. W., ed. 1979. *The Interpretation of Ordinary Landscapes: Geographical Essays.* New York: Oxford University Press.

Migliorino, Nicola. 2008. *(Re)constructing Armenia in Lebanon and Syria: Ethno-cultural Diversity and the State in the Aftermath of a Refugee Crisis.* Oxford: Berghan Books.

Miller, Daniel. 2007. "'Foreword: Getting Behind the Wheel.'" In *The Hummer: Myths and Consumer Culture*, edited by Elaine Cardenas and Ellen Gorman, vii–ix. Lanham, MD: Lexington Books.

Mitchell, Don. 2005. "The S.U.V. Model of Citizenship: Floating Bubbles, Buffer Zones, and the Rise of the 'Purely Atomic' Individual." *Political Geography* 24 (1): 77–100.

Mitchell, Timothy. 1991. "The Limits of the State: Beyond Statist Approaches and Their Critics." *American Political Science Review* 85 (1): 77–96.

Mohamed, Ahmed Amin, and Hadia Hamdy. 2008. "The Stigma of Wasta: The Effect of Wasta on Perceived Competence and Morality." Faculty of Management Technology Working Papers, no. 5, German University in Cairo. http://lss.sub.uni-hamburg.de/vollanzeige?pid=103823.

Mudallali, Amal. 2013. "The Syrian Refugee Crisis Is Pushing Lebanon to the Brink." *Viewpoints*, no. 22, Woodrow Wilson International Center for Scholars. http://www.wilsoncenter.org/publication/the-syrian-refugee-crisis-pushing-lebanon-to-the-brink.

Nagel, Caroline R. 2002. "Reconstructing Space, Re-creating Memory: Sectarian Politics and Urban Redevelopment in Post-war Beirut." *Political Geography* 21 (5): 717–725.

Nagle, John. 2009. "Sites of Social Centrality and Segregation: Lefebvre in Belfast, a 'Divided City.'" *Antipode* 41 (2): 326–347.

Naiden, Fred S., and Kenneth W. Harl. 2009. "Adieu to Lebanon." *Historically Speaking* 10 (2): 24–26.

Nakkash, Tammam, and Neddy Jouzy. 1973. "Beirut Travel Characteristics—A Comparative Study." *Transportation* 2 (4): 411–430.

Nashabe, Omar. 2009. *Security Sector Reform in Lebanon: Internal Security Forces and General Security.* Amman, Jordan: Arab Reform Initiative, Global Facilitation Network for Security Sector Reform. http://www.ssrnetwork.net/doc_library/document_detail.php?id=5691.

Navaro-Yashin, Yael. 2012. *The Make-Believe Space: Affective Geography in a Postwar Polity.* Durham, NC: Duke University Press.

Newcomb, Rachel. 2008. *Women of Fes: Ambiguities of Urban Life in Morocco.* Philadelphia: University of Pennsylvania Press.

Norton, Augustus Richard. 2000. "Hizballah and the Israeli Withdrawal from Southern Lebanon." *Journal of Palestine Studies* 30 (1): 22–35.

———. 2014. *Hezbollah: A Short History.* Princeton, NJ: Princeton University Press.

Notar, Beth E. 2012a. "'Coming Out' to 'Hit the Road': Temporal, Spatial, and Affective Mobilities of Taxi Drivers and Day Trippers in Kunming, China." *City & Society* 24 (3): 281–301.

———. 2012b. "'My Father is Li Gang!': Power and Transgressive Mobility in Contemporary China." Paper presented at an invited session at the American Anthropological Association Annual Meetings, San Francisco, November 16.

Ochs, Juliana. 2011. *Security and Suspicion: An Ethnography of Everyday Life in Israel.* Philadelphia: University of Pennsylvania Press.

Owen, Roger, ed. 1976. *Essays on the Crisis in Lebanon.* London: Ithaca Press.

Packer, Jeremy. 2006. "Becoming Bombs: Mobilizing Mobility in the War of Terror." *Cultural Studies* 20 (4): 378–399.

Pan, Esther. 2006. *Lebanon's Weak Government.* Council on Foreign Relations, July 20. http://www.cfr.org/lebanon/lebanons-weak-government/p11135.

Park, Robert E., Ernest Burgess, and Roderick McKenzie. 1925. *The City.* Chicago: University of Chicago Press.

Peleikis, Anja. 2006. "The Making and Unmaking of Memories: The Case of a Multi-confessional Village in Lebanon." In *Memory and Violence in the Middle East and North Africa,* edited by Ussama Makdisi and Paul A. Silverstein, 133–150. Bloomington: Indiana University Press.

Pellow, Deborah. 2008. *Landlords and Lodgers: Socio-spatial Organization in an Accra Community.* Chicago: University of Chicago Press.

Penglase, R. Ben. 2014. *Living with Insecurity in a Brazilian Favela: Urban Violence and Daily Life.* New Brunswick, NJ: Rutgers University Press.

Perry, Mark. 2002. "Ecological Health Movement in Lebanon: An Overview of Alternative Culture in a Developing Country." *Journal of Ecological Anthropology* 6 (1): 50–67.

Peteet, Julie. 2005. *Landscape of Hope and Despair: Palestinian Refugee Camps.* Philadelphia: University of Pennsylvania Press.

———. Forthcoming. *Space and Mobility in Palestine.* Philadelphia: University of Pennsylvania Press.

Peterson, Mark Allen. 2011. *Connected in Cairo: Growing Up Cosmopolitan in the Modern Middle East.* Bloomington: Indiana University Press.

Picard, Elizabeth. 1996. *Lebanon: A Shattered Country.* New York: Holmes & Meier.

Raad, Walid. 2007. *Let's Be Honest, the Weather Helped: The Raad Files in the Atlas Group Archive.* Cologne: Buchhandlung Walthe König.

Riachi, Georges. 1963. "Beirut." In *The New Metropolis in the Arab World,* edited by Morroe Berger, 103–114. New York: Octagon Books.

Rieker, Martina, and Kamran Asdar Ali. 2008. *Gendering Urban Space in the Middle East, South Asia, and Africa.* New York: Palgrave Macmillan.

Rizk, Carol. 2009. "Baroud Takes Beirut Traffic Nightmare into His Own Hands." *Daily Star,* December 7. http://www.dailystar.com.lb/News/Local-News/Dec/07/Baroud-takes-Beirut-traffic-nightmare-into-his-own-hands.ashx#axzz1v33vF6Bi.

Rojas Pérez, Isaias. 2008. "Writing the Aftermath: Anthropology and Post-conflict." In *A Companion to Latin American Anthropology*, edited by Deborah Poole, 254–275. Malden, MA: Blackwell.

Rotberg, Robert I., ed. 2003. *When States Fail: Causes and Consequences*. Princeton, NJ: Princeton University Press.

Rowe, Peter, and Hashim Sarkis, eds. 1998. *Projecting Beirut: Episodes in the Construction and Reconstruction of a Modern City*. Munich: Prestel Books.

Roy, Ananya, and Nezar AlSayyad, eds. 2003. *Urban Informality: Transnational Perspectives from the Middle East, Latin America, and South Asia*. Lanham, MD: Lexington Books.

Saliba, Robert. 2004. "Looking East, Looking West: Provincial Eclecticism and Cultural Dualism in the Architecture of French Mandate Beirut." In *The British and French Mandates in Comparative Perspective*, ed. Nadine Méouchy and Peter Sluglett, 203–215. Leiden: Brill.

Salibi, Kamal. 1988. *A House of Many Mansions: The History of Lebanon Reconsidered*. London: Tauris.

Sarkis, Hashim. 2014. "Let Beirut Choose History over a Highway." *Daily Star*, April 4. http://www.dailystar.com.lb/Opinion/Commentary/2014/Apr-04/252232-let-beirut-choose-history-over-a-highway.ashx#axzz2zikXxE00.

Sawalha, Aseel. 2003. "'Healing the Wounds of the War': Placing the War-Displaced in Postwar Beirut." In *Wounded Cities: Destruction and Reconstruction in a Globalized World*, edited by Jane Schneider and Ida Susser, 271–291. Oxford: Berg.

———. 2010. *Reconstructing Beirut: Memory and Space in a Postwar Arab City*. Austin: University of Texas Press.

Sbaiti, Nadya. 2009. "'If the Devil Taught French': Strategies of Language and Learning in French Mandate Beirut." In *Trajectories of Education in the Arab World: Legacies and Challenges*, edited by Osama Abi-Mershed, 59–79. Abingdon, UK: Routledge.

Schmid, Heiko. 2002. "The Reconstruction of Downtown Beirut in the Context of Political Geography." *The Arab World Geographer/Le Géographie due Monde Arabe* 5 (4): 232–248.

Schneider, Jane, and Ida Susser, eds. 2003. *Wounded Cities: Destruction and Reconstruction in a Globalized World*. Oxford: Berg.

Scott, James C. 1985. *Weapons of the Weak: Everyday Forms of Resistance*. New Haven, CT: Yale University Press.

———. 1998. *Seeing Like a State: How Certain Schemes to Improve the Human Condition Have Failed*. New Haven, CT: Yale University Press.

Sehnaoui, Nada. 2002. *L'Occidentalisation de la vie quotidienne à Beyrouth, 1860–1914*. Beirut: Éditions Dar an-Nahar.

Sennett, Richard. 1974. *The Fall of Public Man: On the Social Psychology of Capitalism*. New York: Vintage Books.

———. 1994. *Flesh and Stone: The Body and the City in Western Civilization*. New York: Norton.

Sharma, Aradhana, and Akhil Gupta, eds. 2006. *The Anthropology of the State: A Reader*. Malden, MA: Blackwell.

Sikimic, Simona. 2010. "Radar Traps Catch Thousands of Speeders within Hours." *Daily Star*, November 9. http://www.dailystar.com.lb/News/Lebanon-News/2010/Nov-09/59621-radar-traps-catch-thousands-of-speeders-within-hours.ashx.

Singerman, Diane, ed. 2011. *Cairo Contested: Governance, Urban Space, and Global Modernity*. Cairo: American University of Cairo Press.

Singerman, Diane, and Paul Amar, eds. 2006. *Cairo Cosmopolitan: Politics, Culture, and Urban Space in the New Globalized Middle East*. Cairo: American University of Cairo Press.

"The Sixth War: Israel's Invasion of Lebanon." 2006. *MIT Electronic Journal of Middle East Studies* 6 (Summer): 1–233. www.mafhoum.com/press10/292P6.pdf.

Smith, Neil. 2000. "What Happened to Class?" *Environment and Planning A* 32 (6): 1011–1032.

———. 2002. "New Globalism, New Urbanism: Gentrification as Global Urban Strategy." *Antipode* 34 (3): 427–450.

Stevenson, Tala Nasr. 2007. "Informal Consent: The Complexities of Public Participation in Post-Civil War Lebanon." PhD diss., University of Southern California.

Stoller, Paul. 1982. "Signs in the Social Order: Riding a Songhay Bush Taxi." *American Ethnologist* 9 (4): 750–762.

Tahan, Lina. 2005. "Redefining the Lebanese Past." *Museum International* 57 (3): 86–94.

Taxi Ballad. 2012. Directed by Daniel Joseph. Beirut: Lunar Landing Studios.

Taxi Beirut. 2011. Directed by Hady Zaccak. Beirut: Al Jazeera Documentary Channel.

Taxi Talk. 2009. Beirut: Lens on Lebanon.

Theidon, Kimberly. 2013. *Intimate Enemies: Violence and Reconciliation in Peru*. Philadelphia: University of Pennsylvania Press.

Thompson, Elizabeth. 2000. *Colonial Citizens: Republican Rights, Paternal Privilege, and Gender in French Syria and Lebanon*. New York: Columbia University Press.

Tlaiss, Hayfaa, and Saleema Kauser. 2011. "The Importance of 'Wasta' in the Career Success of Middle Eastern Managers." *Journal of European Industrial Training* 35 (5): 467–486.

Traboulsi, Fawwaz. 2007. *A History of Modern Lebanon*. London: Pluto Press.

Totten, Michael J. 2005. "Driving." *World Affairs Journal*, November 20. http://worldaffairsjournal.org/blog/michael-j-totten/driving.

Truitt, Allison. 2008. "On the Back of a Motorbike: Middle-Class Mobility in Ho Chi Minh City, Vietnam." *American Ethnologist* 35 (1): 3–19.

UNDP (United Nations Development Programme). 2011. *State and Trends of the Lebanese Environment: 2010*. 353 pp. http://cskc.daleel-madani.org/resource/state-trends-lebanese-environment.

UN-Habitat. 2012. *The State of Arab Cities 2012/2013*. www.citiesalliance.org/sites/citiesalliance.org/files/SOAC-2012.pdf.

U.S. State Department, Bureau of International Narcotics and Law Enforcement Affairs. 2009. "U.S. Delivers Police Vehicles to the Lebanese International Security Forces." *INL Beat.* www.state.gov/documents/organization/131303.pdf.

Varzi, Roxanne. 2006. *Warring Souls: Youth, Media, and Martyrdom in Post-revolution Iran.* Durham, NC: Duke University Press.

Verdeil, Eric. 2005. "Plans for an Unplanned City: Beirut (1950–2000)." *Worldview: Perspectives on Architecture and Urbanism from around the Globe.* Architectural League of New York. http://www.worldviewcities.org/beirut/urban.html.

Vloeberghs, Ward. 2012. "The Politics of Sacred Space in Downtown Beirut." In *Popular Housing and Urban Land Tenure in the Middle East: Case Studies from Egypt, Syria, Jordan, Lebanon, and Turkey,* edited by Myriam Abasa, Baudouin Dupret, and Eric Denis, 137–168. Cairo: American University of Cairo Press.

Volk, Lucia. 2010. *Memorials and Martyrs in Modern Lebanon.* Bloomington: Indiana University Press.

Vora, Neha. 2013. *Impossible Citizens: Dubai's Indian Diaspora.* Durham, NC: Duke University Press.

Watenpaugh, Heghnar. 2004. "Museums and the Construction of National History in Syria and Lebanon." In *The British and French Mandates in Comparative Perspective,* edited by Nadine Méouchy and Peter Sluglett, 185–202. Leiden: Brill.

Wedeen, Lisa. 1999. *Ambiguities of Domination: Politics, Rhetoric, and Symbols in Contemporary Syria.* Chicago: University of Chicago Press.

West Beirut. 1998. Directed by Ziad Doueiri. Beirut: Doueiri Films.

Wick, Livia. 2011. "The Practice of Waiting under Closure in Palestine." *City & Society* 23 (1): 24–44.

Yahya, Maha. 2007. "'Let the Dead Be Dead: Communal Imaginaries and National Narratives in the Post–Civil War Reconstruction of Beirut." In *Urban Imaginaries: Locating the Modern City,* edited by Alev Çinar and Thomas Bender, 236–266. Minneapolis: University of Minnesota Press.

Yarwood, Richard. 2007. "The Geographies of Policing." *Progress in Human Geography* 31 (4): 447–465.

Zhang, Li. 2010. *In Search of Paradise: Middle-Class Living in a Chinese Metropolis.* Ithaca, NY: Cornell University Press.

INDEX

Page numbers in *italics* refer to figures.

16th Brigade, 161n6
1975 (bar), 155n25

ABC mall, 68, 158n20
Abrams, Philip, 138, 143
Abu Dhabi, United Arab Emirates, 48
Abu Karroum, Ghazi, 79
Abu-Lughod, Lila, 13
Access program, 72
Adnan, Etel, 33, 41
age and security, 89
agriculture, 24, 149n16
Ain-al-Tiné (neighborhood), 95
Ain el Mreisse (neighborhood), 79, 99
air degradation, 33
'Ajat as-sayr (Traffic Jam) (play), 145n5
al-Amin mosque, Mohammad, 59–60
al-Assad, Bashar, 139
Al-Ayoubi, Mohammad, 129, 130, 131, 132–133
Aleppo, Syria, 140, 151n38
Al-Fassad (television show), 117, 145n5
al-Harithy, Howayda, 59
Allied Powers, 24
al-Nour (radio station), 65
al-Nusra Front, 139
Al Qaeda, 139
al-Solh, Riad, 52
Amal Movement (political party), 63, 154–155n22
American University of Beirut, 6, 59, 157n8
amnesia, collective, 47
Anderson, Ben, 88
Anderson, Benedict, 112, 114
Anderson, Terry, 1
An-Nahar (newspaper), 6
anxiety, 3, 6
Aoun, Michel, 97, 154n22
Appadurai, Arjun, 14
Aql, Mona, 110

Aql, Ziad, 110, 117, 145n5
Arab Gulf, 130
Arab-Israeli war, 40, 147n1, 154n17
Arab League, 45, 155n27
Arab Near East, 24
architecture, 26, 150n24
Aretxaga, Begoña, 142
Aridi, Ghazi, 150n27
Armenian Genocide, 21
Ashrafieh (neighborhood), 20–21, 25, 42, 64, 66, 68, 101, 155n28
Association of Automobile Importers, 152n39
automobility: accidents, 102, 142, 159n2; car ownership, 22, 29, 107–108; dependency, 30, 152n39; implications of, 13; public space and—, 29–33; relationship with warfare, 80; security caravans, 83–85. *See also* traffic
The Autostrad: A Mezé Culture—Lebanon and Auto-mobility (Notre Dame University, Louaize), 112

Balkanization, 37
banking industry, 22, 24
Barakat, Halim, 36
Baroud, Ziad, 121–123, 130, 137–138, 143, 161n2
barriers, traffic, 80, 81–85, *84*, 88, 96, 122
Beirut, Lebanon: built environment/physical form, 25–29; Central District (BCD), 148n5; chaos of driving, 101–120; downtown area, 20, 25, 48–49, 155nn30,31; geography, 9, 18, *19*, 20, 42; "growth machine," 33–34, 152n49; insecure city, 8–11, 143–144; physical landscape, 18–34; politics and public space, 56–78; population, 18; port city, 30; privatization, 18–34; redevelopment, 59–60, 148n5; rise of modern Beirut, 21–25; securing, 79–100; space of war, 35–55; tent city, 52, *52*; traffic management, 121–138; urban citizenship, 141

177

Beirutization, 36–37
Bekaa Valley, 24, 39, 47, 61, 151n38
Belfast, Northern Ireland, 51
Berri, Nabih, 63, 95, 96, 154–155n22
Beyhum, Nabil, 50
biblical importance, 148n8
biometric technologies, 88
biopolitics, 88
Black September, 40
blockade (traffic), 2, 10, 12, 80, 95, 96, 140
bomb shelter, 44
bombing, 2–4, 7, 8, 53–54, 79–80. *See also* violence
borders/boundaries, 58–64, 68–77
Bosnia, 51
Bourdieu, Pierre, 114
Brazil, 131
Bronze Age, 35
Burj al-Barajneh refugee camp, 147n1
bus stops, 28
bus transportation, 30, 33, 63–64, 104, 107, 120, 135, 152n43, 155n31, 160n3
Bush, George W., 38
Byzantine era, 35

Cairo, Egypt, 71, 94
Canaanite era, 35
canton, 42
capitalism, 76, 147n19, 152n51
Catholics, 35, 38, 149n14. *See also* Christian; Maronite Catholics
Cedar Revolution, 53
Central Fund for the Displaced, 156n33
Centre d'Etudes et le Recherche sur le Moyen-Orient, 6
chaos, culture of, 111–114
Chamoun, Camille, 38, 39
Chatterji, Roma, 3
checkpoints (traffic), 10, 12, 43, 80, 93, 96–97
Chehab, Fouad, 161n6
Christian: affiliation, 7; landmarks, 59; Lebanese citizenship, 40; neighborhood/community, 2, 24, 36, 39, 40, 42, 56, 58, 66, 68, 70, 82, 98; population, 38, 46, 153n5. *See also* Catholics; Greek Orthodox; Maronite Catholics

citizenship, 14–15, 40, 50, 116, 119–120, 140–142, 144
city planning policy. *See* planning policy
civic life, transformation of, 41–45, 115–116
civilizing mission, 25
clothing, 12, 89
clubbing, 68–73
Comaroff, Jean, 122
Comaroff, John, 122
communications networks, 75
Communist Party, 41–42
community policing, 86
congestion. *See* traffic
constitution, 79
consumerism, 50, 69–70, 74, 76
consumption, practice of, 14
co-presence/face-to-face interaction, 3, 4
Corniche (boardwalk), 31, 32, 47
corporate citizenship, 50
corruption, 116–119, 135–136, 142; elite, 11
Council for Development and Reconstruction (CDR), 49, 117, 150n26
Crusader era, 35
cultural production, 47, 70; global youth culture, 75, 77
Czeglédy, André, 104

Dahiya, Lebanon, 2, 5, 58, 60–63, 157n13
daily life, transformation of, 41–45
Damascus, Syria, 140, 151n38
DaMatta, Roberto, 131
danger, use of term, 11
Deeb, Lara, 9, 16, 61–62, 71, 157n4
de Koning, Anouk, 71
democracy, limits of, 38
deregulation, 11
Directorate General for Urban Planning, 150n26
disorganization, social, 116–119
"the displaced," 156n33
diversity, 36
Document of National Understanding, 46
Doha, Qatar, 48
driving/drivers, 27–28, 143–144; behavior/emotion, 11, 12–13, 65, 113–114, 134–135, 146n17; chaos, 101–120, 142; corruption,

116–119; disorganization, moral and civic propriety, 115–116; education, 109–111, 136; and Lebanese-ness, 111–114; license, 115, 135, 136, 142, 162n16; mobile and differentiated citizens, 119–120; salary, 103; security and—, 80; service taxi, 13, 30–31, 62, 64, 102–105, 160nn4,9; social class and—, 105–109, 134–135; student, 109; traffic violations, 128, 130–131; and urban development, 109–111
Druze, 24, 35, 38
Dubai, United Arab Emirates, 48, 51
Dunes shopping plaza, 69, 158n20

economy, 6, 8, 11, 24, 39, 143, 147n19
education, 6, 43, 147–148n1. *See also* schools
Egypt, 39
electoral reform, 58–59, 157n6
El-Jor, Hanna, 111
employment, 39–40, 140, 147–148n1
enclavization, 93–98, 100
energy hawking, 29
England, 24–26
environmental degradation, 33
errand, running by car, 13, 108
European Union, 130
exterior walls (buildings), 42–43, 86

Fadlallah, Mohammad Hussein, 115–116
family life, 5
famine, 149n13
fare, transportation, 31, 62
fatwa [formal legal opinion or decree], 115–116
favoritism network, 135
Fawaz, Youssef, 27, 89, 93
Fayad, Rahif, 28, 51
feminist thought, 13, 147n19
finance industry, 22, 149n16
fitna [upheaval, disturbance, strife], 116
flags, colors of, 60–61, 64
Flesh and Stone (Sennett), 119
food, 6, 112, 149n13
footwear, 89
forgetfulness, collective, 46
Foucault, Michel, 93
fowda, 115–116

France: influence, 25; invasion of Syria, 24; investments, 22; language, 25; Mandate of Lebanon, 37–38
Frangieh, Suleiman, 155n27
Free Patriotic Movement, 154–155n22
Fregonese, Sara, 81
Future Movement, 60, 63, 83

gas costs, 103
gated community. *See* housing
Gavin, Angus, 50
Geitawai (neighborhood), 32–33
Gemayel, Bashir, 155n28
Gemmayzeh, Lebanon, 27, 70
gender: differences, 13; gendered spaces/ interactions, 14, 147n19, 159n11; inequality, 13–14; mobility entitlement, 89–90, 136; and policing encounter, 162n17; security and—, 86, 88, 89, 90, 128–129, 154n15, 161n9; and service drivers, 104, 160nn4,9
Gharbieh, Ahmad, 89, 93
Gholam, Nabil, 26–27
Giddens, Anthony, 3
globalization, agents of, 13
Goffman, Erving, 90
Golan Heights, 154n17
Graham, Stephen, 94
Greek Catholic, 38
Greek Orthodox, 24, 38
Green Line, 36, 42, 48, 56

Haiti, 127
Hama, Syria, 140
Hamadeh, Marwan, 79, 99
Hammoud, Bourj, 21
Hamra (neighborhood), 71, 155n28
Hannerz, Ulf, 91
Harb, Mona, 9, 16, 61, 71, 89, 93, 157n4
Haret-Hreik (neighborhood), 28, 29
Hariri, Rafiq: assassination, 2, 3, 10, 48, 52, 57, 60, 67, 73, 81–82, 87, 88, 94–95, 99; burial site, 52, 53; LCC ownership, 63; radio station, 65; residence, 93–94, 97; resignation, 7; urban development projects, 48–49, 51, 59

Haugbolle, Sune, 9, 47, 65
Haussmann, Baron de, 150n21; Haussmannian Paris, 25
Hazmieh, Lebanon, 62
headscarf, 89
health-related insecurity, 143, 147–148n1
helicopters, 94
Hellenistic era, 35
helmet law, 121, 143
Helou, Elie, 83, 118
Hermel, Lebanon, 162n1
Higher Council for Urban Planning, 150n26
Hizbullah, 5, 54, 115, 145n1, 147n1, 162n1; capture of Israeli soldiers, 53; flag, 60; growth of, 123; participation in Syrian civil war, 140; radio station, 65; rebuilding of Haret-Hreik neighborhood, 28, 29; security forces, 62; support for, 44–45, 57, 78, 157n3; war outside Lebanon, 139; war with Israel, 8, 98
Holiday Inn, 48, 54
Holston, James, 14
Homs, Syria, 140
Horsh (forest), 31
hospitals, 25
housing, 18, 19, 147n19; balcony, 18, 20; gated community, 81, 86, 94, 147n19; high-rise towers, 31; production, 28; rents, 143; security, 43–44, 86; settlements, 28–29
Human Rights Watch, 2
Hussein, Imam, 58
"hybrid" sovereignty, 81

iconography, 60
infrastructure, 9, 24, 27
insecurity, 7, 9, 10, 11. *See also* security/safety
insurance industry, 22
interior design, 147n19
Internal Security Forces (ISF), 15–16, *15*, 121–123, 127–128, 129, 130, 138, 161nn1,10
intersectarian tension, 2
Iran, 44, 56, 130
Iraq, U.S. invasion, 38
ISF. *See* Internal Security Forces
Islamic extremists, 139

Israel: founding of, 39, 40, 147n1; ground and air strikes, 2; invasion of Lebanon, 44, 142, 145n1; occupation, 12, 36, 52, 145n1, 154n17; resistance to, 60; soldiers captured, 53; war with Hizbullah, 8, 98
Israel Defense Force, 2

Jegnathan, Pradeep, 67
Jesuit Garden, 32
Johannesburg, South Africa, 104
Jordan, 40
Joseph, Suad, 39–40
Jumblatt, Walid, 150n27
Jyllands-Posten (Danish newspaper), 65–66

Karantina, Lebanon, 21
Katz, Jack, 12
Khabbaz, Ilham, 106, 145n4
Koreitem (neighborhood), 10, 93–94

Lahoud, Emile, 7, 54, 79
laissez-faire, state of, 148n9
language: Arabic accents, 89; class marker, 12; French, 25, 109, 150n23; "Lebanize" Arabic, 4; of service taxi drivers, 104
Lebanese Commuting Company (LCC), 63
Lebanese National Archives, 6
Lebanese-ness, 111–114
Lebanese Phalange, 40–41, 65, 155n28, 157n17
Lebanese Transparency Association, 117
Lebanon: Bureau of Intelligence, 128; citizens and state, 14–15; citizenship, 40; civil war, 1, 36–37, 145n6, 153n3; constitution, 79; creation of state, 24; culture, 112; geography, 148n5, 149n11; government, 24–25, 38, 46; Interior Ministry, 128; population, 38; regional disparities, 24; regional war, 1; state of, 16, 121–138
Lefebvre, Henri, 78, 143, 152n51, 161n17
leisure sites/practices, 9, 22, 23, 57, 62, 70–71, 157n4
Levant, 25
license plate, 136–137, 140
lifestyle practices, 69–70
locust plague, 149n13
L'Orient le Jour (newspaper), 6

Majal (urban planning institute), 27
Makdisi, Jean, 53, 67
Makdisi, Saree, 46, 155n30
Mamluk era, 35
Mandate of Lebanon, 37–38
Mansel, Philip, 24
manufacturing, 21
March 8 coalition, 7, 53, 56–57, 62, 63, 65, 68, 77, 78
March 14 coalition, 2, 7, 53, 54, 56, 57, 60, 62, 63, 65, 68, 77, 82, 87, 157n17
Maronite Catholics, 24, 38, 39, 40–41, 61, 149nn14, 15, 157n17. *See also* Catholics; Christian
Martyrs' Square, 52–53, 57, 155n31
mass media, 75–76
Mathaf-Barbir Crossing, 36
Mecca, 58
Mehta, Deepak, 3
memorial/monument, 9, 60
memory culture, 46
Meouchi, Badri, 117–118
migrants, 73; rural-to-urban, 22
Mikati, Najib, 128
militia, 41, 44–45, 46, 48, 153n10, 154n15
minibus transport, 30, 152n43
Ministry of Displaced, 156n33
Ministry of Public Works and Transport, 30, 150n26
mobility, 1–2; citizenship, 140–142; entitlement, 89–90; experiences, 14; field of mobility studies, 146n18; geographic, 72, 76; impacted by security, 80; skills and tactics, 91; socioeconomic, 11, 145nn3, 7; spatial, 8, 13; urban space and—, 12–14. *See also specific transportation mode*
Mohammad, Prophet, 2, 57, 65–66
Mohieddine, Ali, 105
Monot (nightclub areas), 70
morality, 9, 115–116, 157n4
Mostar Bridge, 51
motor scooters, 87–88, 90, 143
Mount Lebanon region, 21, 22, 24
multiagency policing, 81
"multiple memory cultures," 65

Muslim: concept of, 37; landmarks, 59; neighborhood/community, 36, 40, 42, 56, 66; population, 38, 46, 149n15, 153n5; service drivers, 64. *See also* Shiʿi Muslim; Sunni Muslim

Nasrallah, Hassan, 52, 139, 162n1
Nasser, Gamal Abdel, 39
national origin, 89, 90
National Museum of Beirut, 35–36
National Pact, 38
neighborhood, 18, 20, 20–21, 56, 58, 62, 71, 157n5. *See also specific neighborhood*
neighborhood watch groups, 86
newspapers, 65
Niger, 105
noise, 30, 33, 125

"the occupiers," 156n33
OGER Liban, 49
oil pipeline, 30
oligarchy, 39
open market, 11
Ottoman Empire, 21, 24, 51, 150n23

Palestine, 12, 18, 36, 39, 40–41, 44, 147–148n1, 154–155n22
Palestinian Liberation Organization (PLO), 40
Paris, France, 150n21
parking, 30–31, 97, 108, 125
parking meters, 117–118, 133
parks, 31–33
"Parties and Colors" albums and stickers, 60, 61
paternalistic power, 24, 149n19
patronage, 4, 135–136, 149n15
pedestrian bridges, 117, 118
pedestrians, 27, 80, 82–83, 95, 115, 143–144
Persian era, 35
personal-status laws, 145–146n8
Peterson, Mark Allen, 71
Phalange Party. *See* Lebanese Phalange
Phoenicia Hotel, 48, 52
Phoenician era, 35
Picard, Elizabeth, 40
piety, 61–62, 157n4

pilgrimage, 58
Place de l'Etoile, 25, 49, 150n22
planning policy. *See* urban planning policy
plural policing, 81
police tape, 82, 94
policing agencies, 86, 126, 127–131, 133–134, 161nn6,8
politics: discussions of, 64–65; identity, 58, 145–146n8; public space and—, 56–78; representation, 58; sectarian-based, 24, 58–64, 141, 143–144, 145–146n8; structure, 8, 14
pollution, 143
port modernization, 24
postal service, 24
presidential term, 7, 79
prisoner exchange, 8, 146n9
privatization, 11, 18–21; automobility and public space, 29–33, 148n6; "growth machine," 33–34; lack of planning and informality, 25–29; rise of modern Beirut, 21–25, 48
profiling practices, 87–90
public safety, 13
public space. *See* urban space
public transportation, 7, 13, 29–30, 63–64, 104, 145n4, 152n43; fares, 31, 62, 103; social class and—, 106–107; unpredictability, 28. *See also specific mode;* traffic

Rabieh (neighborhood), 97
race, 13, 86, 88
Radio Orient, 65
radio stations, 65
rail transport, 29–30, 151n38
Raouché (property), 47
real estate development, 48
Reconstructing Beirut: Memory and Space in a Postwar City (Sawalha), 8
recreation. *See* leisure sites/practices
refugee/displaced person, 2, 54, 61, 139–140; camp, 18, 40, 147–148n1, 154n12, 154–155n22
religion, 145–146n8; communities, 24; scripture, 115–116; and violence, 22, 149n11. *See also specific religion*
Riad al-Solh (square), 57

Rifi, Ashraf, 123, 128, 138
rioting, 67
risk, management of, 11
roads: building, 22, 118; closure, 12, 97–98, 122, 135, 140–141; configuration, 96; names, 27; narrow, 30; network, 24; safety, 121–123: signage, 27; walking in, 115. *See also* traffic
Rojas-Pérez, Isaias, 54
Roman era, 35

Sabra, Lebanon, 44
Sabra-Shatila refugee camp, 147n1
safety. *See* security/safety
Safina Group, 60
Saida, Lebanon, 40
Saint Joseph's University, 6
Saliba, Robert, 26
Sanayeh Garden, 66
sanitary services, 25
São Paulo, Brazil, 94
Sarkis, Hashim, 33
Sawalha, Aseel, 8–9, 50
Sbaiti, Nadya, 150n23
schools, 2, 25, 43. *See also* education
Scott, James, 124
seatbelts, 130
sectarianism, 8; breakdown of trust, 44; conflict, 37; cross-sectarian community solidarity, 9; culture of, 149n11; identity, 41–43, 58–59, 145n8; political, 14–15, 58–64; quotas, 46; "sectarianness," 63–64; security and—, 89; segregation, 51; territorialization, 141
security/safety, 79–100, 128; barriers, 81–85; caravans, 83–85; enclavization, 93–98, 100; encounter, 90–93, 162n17; forms of, 62, 144, 158n24; homegrown, 86; impact on security, 80; inequality, 99–100; installations, 80–83; intensification, 158n1; measures, 2 (*see also specific measure*); navigating, 90–93; notions of, 69, 73, 75, 141; private-public, 81, 85–87, 94; profiling practices, 87–90; social class and—, 98–99; surveillance, 81, 86, 89; tactics, strategies, competence, 88, 90–93; theories, 11

security guards. *See* security
"securocracy," 81, 159n5
segregation, racial/class, 13
Sennett, Richard, 3, 119
Senioura, Fouad, 91–92
Serof, Gregoire, 31
service-oriented economy, 39
setback, 27
settlements, 28–29
Shankaboot (web-based video series), 145n5
Shatah, Mohammad, 54
Shatila, Lebanon, 44
Shiʻi Muslim, 7, 9, 24, 29, 38, 39, 41, 44–45, 58, 61, 63, 115, 128
shipping industry, 22, 30, 149n13
silk industry, 22
Sitt Marie Rose (Adnan), 33, 41
snipers, 43
social behavior/practice, 9
social class, 1, 8, 11, 13, 72, 140; differences, 13; driving and—, 105–109, 134–135; exclusion, 51, 87; hierarchies, 120, 131; inequality, 8, 9, 99–100, 141–142, 145nn3,8; license plate and—, 137; mobility and—, 68–73, 89–90, 136, 139–143; perceptions of violence, 86; police and—, 133–134; positioning, 77; privilege, 77, 133, 141; security and—, 86, 88, 89, 90, 98–99; segregation, 94; solidarity, 39; visible marker, 12, 89
social differences, 8
social disorganization, 116–119
social identity, 90
social inequality. *See* social class
social media, 7
social security system, 38
social services, 147–148n1, 158n24
socialization, 68–73
socioeconomic mobility. *See under* mobility
Sodeco (intersection), 134
Solidere, 4, 151n28; creation of, 49; development projects, 26, 51, 59, 148n5, 156n33; Urban Development Division, 50–51
Solh, Amira, 26
Soviet influence, 39
spatial life, transformation of, 41–45

spatial mobility. *See under* mobility
spatial polarization, 56, 87, 147–148n1
spatial segregation, 94
spatial story, 56, 156n1
speed bumps, 95
Sri Lanka, workers from, 5
state, 14–15; encountering, 142–143; *ma fi dowla*, 17, 124, 127, 142; force of, 129–133; ill-functioning, 125–127; power, 8, 16; production of, 121–138; spatializing, 15–16; traffic management and—, 121–123; verticalization of authority, 132; weak, 123–124
sticker/sticker album, 60
Stoller, Paul, 105
Stone Age, 35
streets. *See* roads
stress. *See* anxiety
Suleiman, Michel, 122
Sunni Muslim, 7, 24, 38, 59, 128
SUVs, 84–85, 159n6
Syrian: anti-Syrian coalition, 2, 56; civil war, 15, 139–140, 142, 162n1; French invasion, 24; influence, 47–48; laborers, 29, 73, 87–88, 107; occupation, 36, 44, 65, 79, 87, 155n27; refugees, 54, 139–140; withdrawal from Lebanon, 2, 7, 52, 52, 68, 78
Syrian Social Nationalist Party, 155n28

Taʾif, Saudi Arabia, 46
Taʾif Accord, 46, 59, 154n20
Tanzimat reforms, 148n10
Tariq al-Jedeideh (neighborhood), 67
taxes, 6
Taxi Beirut (documentary film), 145n5
Taxi Drivers Syndicate, 105, 145n4
Taxi Talk (documentary film), 145n5
taxis: competition, 106; fares, 103; flagging/hailing, 103, 104, 145n4, 160n3; illegal, 105–106, 152n43; license plate, 137; private, 30, 63; service, 7, 30, 62, 63–65, 102–109, 120, 141, 145n4, 152n43, 155n31, 160nn4,9; unregistered, 140. *See also* driving/drivers; public transportation; traffic
telephone network, 24–25

television programs/stations, 7, 65, 117, 145n5
tent city, 52, 52
theater/stage, 7, 145n5
threat level, 89, 93
Totten, Michael, 114
tourism, 30, 48, 50, 51, 130, 152n39, 162n11
Tourist Brigade, 30
trade, 24, 25, 35, 149n16
traffic: barriers, 81–85, 84, 88, 96; commercial, 20; congestion, 30–31, 103, 140, 142, 152n43; encounters, 133–138; hand signals, 134; infrastructure, 27; issue of, 6–16; laws/rules, 13, 115–116, 122, 126; lights, 117, 118, 162n13; management technologies, 129–130; patterns, 27; police, 127–131, 133–134, 143 (*see also* policing agencies); rerouting, 10, 80; safety, 110, 121–123, 130; signs, 77–78, 77; speed, 31; training, 127–129, 161n8; violations/tickets, 128, 130–132, 136. *See also* driving/drivers; public transportation; taxi
tramway, 29, 33
Trans Arabian oil pipeline, 30
Tripoli, Lebanon, 18, 40, 139
tsunami, 5
Tyre, Lebanon, 40, 154n17

United Arab Republic, 39
United Nations Relief and Work Agency (UNRWA), 147n1
United Nations resolution 1559, 78
United States: Bureau of International Narcotics and Law Enforcement, 129; civil war involvement, 36; invasion of Iraq, 38, 44; missionary schools, 25; support from, 7, 39, 45, 56, 129, 145n1
unsafety, use of term, 10, 11
urban citizenship. *See* citizenship
urban culture, 18, 20–21
urbanization, process of, 152n51
urban planning policy, 9, 25–29, 150n26, 151n34; French influence, 25; security, 83; system of, 26

urban space, 1–2; automobility and—, 29–33; care of, 115; fight for, 6; memorials and—, 9; morality and—, 9, 12–14; political sectarianism and boundaries, 58–64; politics and public space, 56–78; redevelopment, 59–60, 109–111; right to, 152n51; social inequality and—, 8, 141
Urban Transportation Development Project (UTDP), 117–118

veiling, practice of, 14
Verdun (street), 95, 158n20
vigilante groups, 81
violence, 2–3, 8, 22, 64–68. *See also* bombing
Virgin Mary, 58
visa, inaccessibility of, 51
Vloeberghs, Ward, 57
Voice of Lebanon (radio station), 65
Volk, Lucia, 9, 60
vulnerability, sense of, 10

Waʿad (private development agency), 28
war, space of, 35–37; context, 37–41; rebuilding, 45–47; transformation of civic, spatial, daily life, 41–45
war crimes, 46
War of the Camps, 154–155n22
wasta [connections], 135–136
Watenpaugh, Heghnar, 35
welfare allocation, 59
World Affairs Journal, 114
World Bank, 117
World Cup, 125
World Press Photo Award, 158n22
World War I, 24

youth activism, 158n26
Youth Association for Social Awareness (YASA), 110, 115, 128–129, 145n5

Zarif (neighborhood), 66
zig-zag traffic barrier, 81–85, 88
zoning, 28, 151n35

ABOUT THE AUTHOR

KRISTIN V. MONROE is an assistant professor of anthropology at the University of Kentucky. Her work has been published in *City & Society, Anthropology of Work Review, Comparative Studies of South Asia, Africa and the Middle East,* and *Everyday Life in the Muslim Middle East,* 3rd edition.

www.ingramcontent.com/pod-product-compliance
Ingram Content Group UK Ltd.
Pitfield, Milton Keynes, MK11 3LW, UK
UKHW021312180426
11947UKWH00015B/1180